# EXTERIOR FINISHES

FROM THE EDITORS OF **Fine Homebuilding**®

The Taunton Press

The Taunton Press
Inspiration for hands-on living®

The Taunton Press, Inc., 63 South Main Street, PO Box 5506, Newtown, CT 06470-5506
e-mail: tp@taunton.com

Distributed by Publishers Group West

Cover Design: Cathy Cassidy
Interior Design: Cathy Cassidy
Layout: Carol Petro
Front Cover Photographer: Andy Engel, courtesy *Fine Homebuilding,* © The Taunton Press, Inc.
Back Cover Photographer: (clockwise from top left) Steve Culpepper, courtesy *Fine Homebuilding,* © The Taunton Press, Inc.;
Andrew Wormer, courtesy *Fine Homebuilding,* © The Taunton Press, Inc.; Tom O'Brien and David Ericson, courtesy *Fine Homebuilding,* © The Taunton Press, Inc.; Roe A. Osborn, courtesy *Fine Homebuilding,* © The Taunton Press, Inc.

For Pros By Pros® is a trademark of The Taunton Press, Inc.,
registered in the U.S. Patent and Trademark Office.

Library of Congress Cataloging-in-Publication Data

Exterior siding, trim, and finishes / the editors of Fine homebuilding.
    p. cm. -- (Taunton's for pros by pros)
Includes index.
 ISBN 1-56158-652-8
 1. Siding (Building materials) I. Fine homebuilding. II. Series: For
pros, by pros.
TH2238.5 .E95 2004
698--dc22

                    2003018412

Printed in the United States of America
10 9 8 7 6 5 4 3 2 1

The following manufacturers/names appearing in *Exterior Finishes* are trademarks:
3M®, Bituthene® 3000, Bondo®, Cabot's® Problem Solver, Cedar Breather®, Cedar Impressions®, Celotex®, Cemplank®, ChemRex®, Colorlok®, ConServ Epoxy®, Cor-A-Vent ®, Cultured Stone®, Dow Chemical, Dow Corning®, Dryvit's® EIFS, Durock® Insulated System, Enkamat®, Exxon, Fiberbond®,Frandon Lead Alert™ Kit, Gold Bond® Jumbo,Gougeon® Brothers, Grace Ice & Water Shield®, James Hardie®, Hitachi® C8FB, Hohmann & Barnard Inc.®, Inner-Seal®, Makita®, Metabo®, Metabo® LF714 S, Midget Louvers®, Mortar Break®, Mortar Net™, Multi Strip™, Paint Shaver®, Penofin®, Plexiglas®, Pratt and Lambert's® Accolade®, Pratt and Lambert® Suprime® 8, ProPrep®, Scotchgard™, Sikaflex®, Snapper™, Stanley Surform®, Sto® Corp, Thompson-Minwax® Co., Typar®, Tyvek®, W.W. Grainger®, Wascoseal®, Wind-Lock™ fasteners.

Special thanks to the authors, editors, art directors,
copy editors, and other staff members of *Fine Homebuilding*
who contributed to the development of the articles in this book.

# CONTENTS

## PART 3: SHEATHING AND SIDING

## PART 4: WOOD REPAIR AND PAINT

# INTRODUCTION

I successfully avoided vinyl siding for the first eight years of my career as a carpenter. By that point, I was pretty good at my job. Not a master by any means, but I could frame roofs, lay out stairs and build custom cabinets. So when the boss asked me to side a new house with vinyl, I did what good carpenters do: I cut the pieces exactly and nailed them up tight. A job well done ... or so I thought until a few months later when the summer sun hit the house.

The siding pooched and dimpled like the surface of golf ball. Turns out that vinyl has what's called a high coeffecient of expansion, which means it expands and contracts a lot with changes in the weather. You're supposed to cut the pieces a little short and not sink the nails completely. All the vinyl on that house had to be torn off and reinstalled.

Interior trim and finishes have to look good, but that's about it. Outside, it's another story. Exterior work has to look good and withstand the rigors of wind-blown rain, blistering sun and freezing temperatures. It isn't work to be taken lightly—a lesson I learned the hard way. You'll want good information about how to do things, and that's where this book can help.

Collected here are 15 articles from past issues of *Fine Homebuilding* magazine. Written by builders from all over the country, these articles represent the hard-won knowledge of professionals who have learned from their mistakes. There's even an article on installing vinyl siding. Wish I'd read that 20 years ago.

—Kevin Ireton,
editor-in-chief, *Fine Homebuilding*

# Running Exterior Trim

■ BY ROBERT WEATHERALL

No part of building a house is more fun than putting up the exterior trim. It's like decorating a cake or coloring in a picture. After the rush of framing lumber and plywood, the trim is the welcome stage of finer joinery that can bring out the house's character.

In more than 17 years of building in New England, I've learned a few things about wood, weather, and details that make trim last. Some lessons come from my own mistakes, but others come from working on old houses and seeing the legacy of carpenters whose work still looks good after more than a century. I like to imagine that the work my crew and I have done will be as praiseworthy a century from now.

## Good Trim Begins with Top-Notch Stock

The first step we take to help ensure that the trim work stays straight and true for years to come is choosing the right material. For exterior trim, we use eastern white pine, D-select or better, almost exclusively.

Western red cedar is also an excellent choice. Cedar is far more rot resistant than pine, and because it's typically quartersawn or vertical grain, there is little chance of cupping and warping. However, cedar is light and brittle and tends to split more easily than pine. Perhaps the best argument for pine is that it's less expensive, and it does not require the harvesting of old-growth trees.

For this house, we used 1-in. select pine (¾-in. nominal thickness) for all the trim except the water tables, which required thicker stock. The water table was made of 2-in. #2 pine that we were fortunate enough to find at a local sawmill. It was far less expensive than the alternative, 8/4 sugar pine.

## Prime Exterior Trim on All Sides

Another step essential to exterior-trim longevity is sealing each piece front and back with primer or stain before it's installed. Moisture can enter trim through any unsealed surface, and sealing all exposed wood surfaces not only helps keep trim from warping and cupping but also keeps moisture from lifting off the finish.

With this project—as with all projects—we try to have the wood on site far enough in advance so that the primer can dry thoroughly before we need to use it, especially

**Exterior trim** may not be scrutinized like interior trim, but it still has to look good and stand up to weather.

in colder weather. But before we prime, we try to make any rip cuts, rabbets, or grooves, such as the groove for the soffit-to-fascia joint. Cutting primed wood—especially when oil-based primer is used—can be hard on sawblades and edge tools. Last, we keep several yogurt containers on hand as primer buckets, and we judiciously prime all exposed wood after every cut.

# Eave Details
# Start with a Drawing

We are always eager to get the roof on our projects, and the roof can't be put on until the eave trim is done. The lines of fascia, soffit, and frieze beneath them also dictate the height of the corner boards and the line of the water table. Window-casing heights are often governed by the line of the eave trim as well.

Before we start cutting and nailing and often before framing, we make a full-scale section drawing of the eave trim. The traditional boxed eaves on this project were more complex than most, done in a vernacular variation of the Greek-revival style. Eave venting was incorporated in the frieze (see drawing below).

The soffit and fascia ended at the bottom edge of the roof with what we call poor-man's crown molding, or flat molding in-

## Stealth Venting for a Traditional Boxed Eave

Ordinary strip or button-type soffit vents can ruin the looks of a complex eave detail, such as the one shown here. Instead, the frieze board is held off the wall sheathing with spacer blocks, and a hidden vent strip allows air to enter and circulate into the rafter bays. Special blocking attached to the rafter tails (inset photo) provides nailing for the inside edge of the soffit and for the flat crown molding.

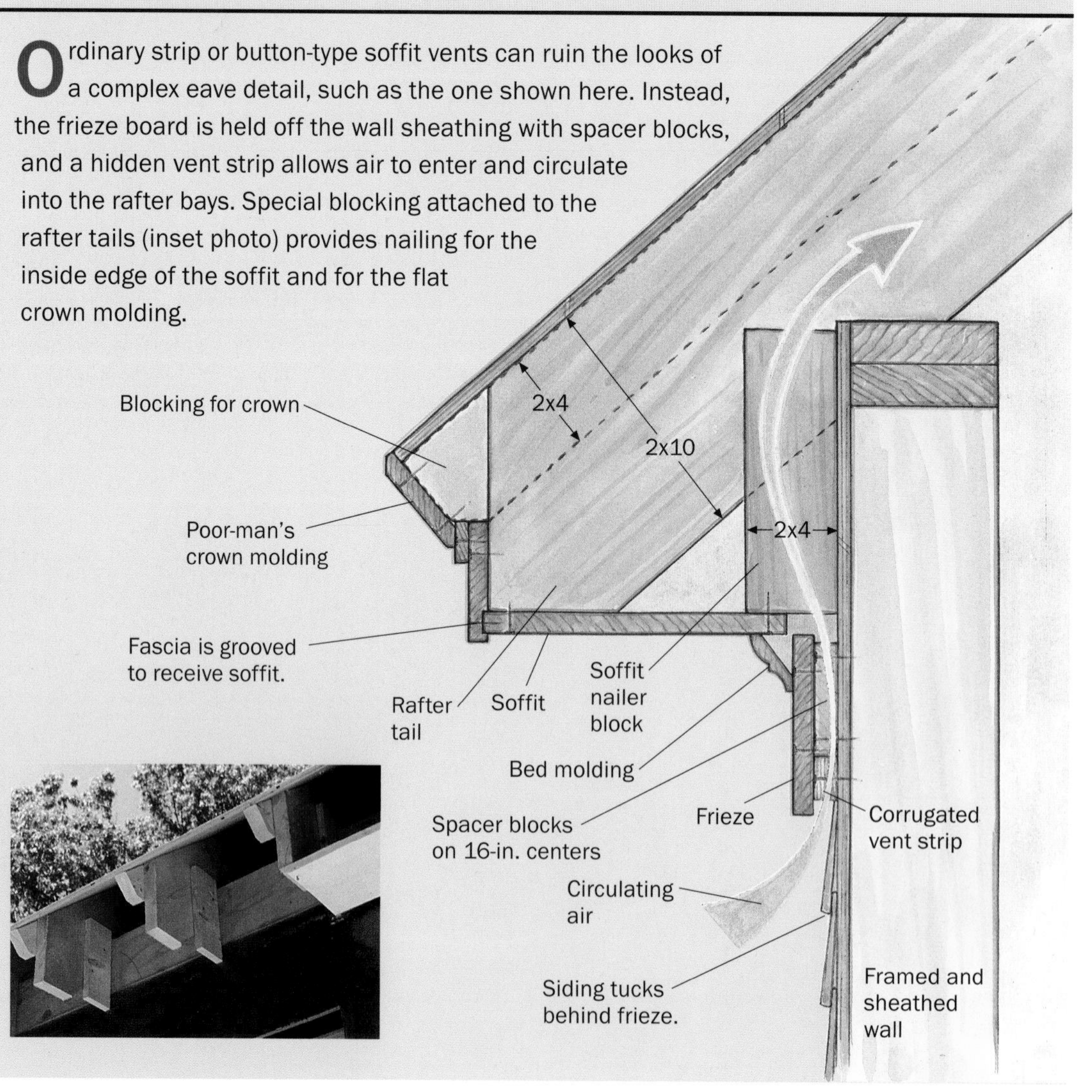

Blocking for crown

Poor-man's crown molding

Fascia is grooved to receive soffit.

2x4

2x10

2x4

Rafter tail

Soffit

Soffit nailer block

Bed molding

Spacer blocks on 16-in. centers

Frieze

Corrugated vent strip

Circulating air

Siding tucks behind frieze.

Framed and sheathed wall

# Building a Cornice Return

On this house, the eave details return, or wrap around the corners at the gable ends. We mitered the ends of the trim where they met the gable returns. First, the blocking was fashioned to support all the layers of the fascia and the crown detail (see top left photo). Because there was no overhang on the gable ends of the house, the soffit detail was eliminated on the return. We nailed in the various layers of the house's trim (see bottom left photo), and then we topped off the return with a layer of lead flashing that was molded into place with a hammer and a wooden block (see photo below).

**All the eave details** except the soffit wrap around onto the gable end of this house. First, solid blocking is installed for attaching the return trim.

After the return trim is installed, a lead cap is molded over the return to keep the rain off. A wood block and a hammer are used to shape lead flashing over the return trim.

**Next, each layer of eave trim** on the return is mitered into the main run and nailed in place.

stalled at a 45° angle. The drawing gave us the dimensions and locations of all the trim elements, as well as the size of the blocking that was to hold the crown and the inside edge of the soffit.

## Fascia and Soffit Are Assembled on the Ground

The fascia on this house overhangs the soffit by ⅜ in. This overhang creates a shadowline, or reveal, along the corner of the eave, which helps hide inaccuracies in the rafters. We created the soffit-to-fascia joint by milling a groove on the backside of the fascia ¼ in. deep and just slightly wider than the ¾-in. thickness of the soffit stock for an easy fit.

We assembled the sections of fascia and soffit on sawhorses, holding the soffit stock vertically and tapping the groove in the fascia down over edge of the soffit (see top right photo). The lengths of the boards correspond to the rafter layout so that when installed, the ends of the boards land on framing. We drive 6d galvanized box nails or stainless-steel siding nails through the fascia to hold the joint together.

Our rule of thumb is to stagger butt joints in long lengths of trim by at least 32 in. (assuming 16-in. rafter layout). With staggered joints, the work runs straighter across the building, and there are no eye-catching areas where a lot of trim ends in the same place.

Rather than using square butt joints, we join pieces for long runs with a bevel or scarf joint. A beveled joint stays closed better despite board movement, and the joint is easier to sand or plane smooth.

When the first section of fascia and soffit was assembled, we tacked it in place (see bottom right photo). We then took measurements for the next section directly from the first section. That section was then cut,

The fascia with a groove cut into the back face fits over the edge of the soffit and is nailed.

The assembled box with the lengths staggered on the rafter layout is lifted and tacked in place until the entire run is in position.

assembled, and tacked in place. When the run was tacked up, we sighted down the edges and faces to make sure they were straight (see photo on facing page). High spots were tapped in with a hammer, and low spots were shimmed out when necessary.

After the fascia and soffit were straightened and nailed off with 8d galvanized box nails, we ran a small rectangular molding that forms a shadowline below the crown. To set the height of the molding, I set my combination square to the right distance and made marks every couple of feet along the fascia. Our poor-man's crown came next, and again, we tacked the length in place, straightening it before nailing it home with stainless-steel siding nails. The ring shank on the siding nails has better holding for smaller molding pieces.

**Eyeballing for straightness.** After each trim element is tacked up, in this case the flat crown molding, it is sighted to make sure that it's straight before the nails are driven home.

**Frieze completes the eave trim.** A wide board called the frieze is installed below the soffit (top). A gauge block is then used to mark the position of the bed molding (center), which covers the joint between soffit and frieze (bottom).

**Measuring down for the water table.** Once the soffits are in, measurements are taken for the height of the water table, which has to run parallel to the soffits for the siding to go on properly.

# Building a Frieze That Breathes

The next molding we installed was the frieze, or the wide flat molding that runs below the soffit. We lay out the frieze by first focusing on the doors and windows just beneath it. On this house, the frieze was to replace the head casings. If there had been space between the head casing and the frieze, we would have sized and located the frieze according to the siding exposure. When the lower edge of the frieze had been determined, a chalkline was snapped along the wall. Because chalklines tend to sag over long distances, we check the accuracy at several places before snapping. At this point, we were not concerned about the gap between frieze and soffit, which was to be covered later by bed molding.

We usually nail 1x3 furring strips beneath the frieze board, which creates a space to hide the top edge of the siding. On this house, however, we were integrating our eave ventilation into the frieze (see drawing on p. 6), so we couldn't use a solid spacer. First, we nailed 1-in.-thick blocks vertically at every stud location. Next, a corrugated vent strip from Cor-A-Vent was nailed in below the blocks. Solid blocking replaced the vent strip at all window and door locations. With the frieze boards nailed on top, this system provided both the eave ventilation and a space for the siding.

Next, the frieze boards were cut and installed over the blocks, and the frieze was returned onto the gable end (see top left photo on facing page). We cut a block of wood to the same dimensions as the bed molding and used it to mark the position of the molding every few feet along the soffit (see middle left photo on facing page). The bed molding was then nailed in following the lines (see bottom left photo on facing page) but was stopped at the corner instead of continuing onto the gable end.

# Water Tables Follow the Soffit Lines

Once the soffits were permanently installed, we were able to measure down for the location of the water table (see photo at right on facing page), which is a wide bevel-edge board that runs along the lower edge of the sheathing. Earlier, we had beveled the top edge of each board with a table saw and a jack plane and then primed the boards.

After the chalklines are snapped for the water table, we miter the ends of a couple of cut-off scraps. As we work our way around the house, we use the scraps to test the corners, which can be a bit out of square. Once we've determined the exact angle at each corner, the scrap is left tacked in place and used for pulling our measurements (see photo at top on p. 12).

Next, we cut the boards and tack them in place using the scraps to determine the fit (see bottom left photo on p. 12). Long pieces are joined together with 45° scarf joints. We used a block plane to tune the scarf joint wherever needed. When we were satisfied with the straightness and fit of a length of water table, we nailed it off with 10d galvanized box nails and moved to the next side of the house. Lead flashing was added for the corner boards (see bottom right photo on p. 12), and later, when we sided the house, the rest of the water table was flashed with the flashing tucked under the housewrap for proper water shedding.

## TIP

*Chalklines tend to sag over long distances, check the accuracy at several points before snapping.*

**Scraps help with measuring and fitting.** Cut-off scraps of water table are tacked on the corners to test for square. The scraps are then left in place to help measure lengths.

## Sources

**Cor-A-Vent®**
(800) 837-8368
www.cor-a-vent.com

After the water table has a test-fit, the boards are nailed in place.

**Lead flashing** for the bottom edges of the corner boards is slipped behind the housewrap and over the water table.

# Buttoning Up the Corners

Corner boards connect the eave trim with the water table. This house was to be sided with cedar clapboards, so we were able to use ¾-in. stock for the corners. If a house is sided with shingles, we use 5/4 stock that can hide the edges of the shingle butts. If 5/4 stock is used for corner boards, we use the same thickness for window and door casing.

As with the eave, we preassembled the sections of corners on the ground before putting them up (see photo at right). Pre-assembling the corner sections helps them span any irregularities in the sheathed corners of the house. And it's much more difficult to achieve a tight corner joint when assembling corner boards in place.

To make the corner boards the same size in each direction, we ripped an amount equal to the thickness of the stock off one of the two legs. Then we measured, cut, and nailed the boards together on the sawhorses.

We were able to build these corners with single boards, but when the corner height is greater than the stock length (for a two-story corner, for instance), we make up the required length by joining the lengths of stock with beveled scarf joints. These joints in the corner boards have to be cut so that water sheds to the outside of the boards. We try not to make any piece of corner-board trim less than 3 ft. long, and we offset the scarf joints on the two legs. For example, if 5 ft. is added to the bottom of a 16-ft. piece on one leg, we put the 5-ft. section at the top of the adjacent leg.

We slid the assembled corners up under the frieze boards and again tacked them in place to assess their straightness and squareness. When we were satisfied, we nailed off the corners, driving most of the nails along the outer edges of the boards (see top photo

**Finishing off the corners.** Corner boards are nailed together on the ground before they go on the house.

on p. 14) with an occasional nail driven close to the corner where the boards were joined. Nails should not be driven in the middle of the boards where they can split.

The final detail we applied was the capital trim around the tops of the corner boards (see bottom photo on p. 14). Trim pieces fit under flashing that had been slipped under the frieze before the corners were installed.

> *Preassembling the corner sections helps them span any irregularities in the sheathed corners of the house.*

Preassembled corner boards span irregularities in the framing and sheathing more easily. Nails driven along the edges of the corner boards are less likely to split the wood.

For that final decorative note, a small band of capital molding is popped in under a drip cap to finish off the tops of the corner boards.

# Rake Trim Finishes off the Gables

We usually duplicate the soffit and fascia detail on the rakes, but without gable over- hangs, the best we could do was to make the rakes mimic those details (see drawing on facing page). We started by running furring- strip spacers that hold the main rake board ¾ in. off the face of the gable wall. Just as with the frieze boards, the furring strips gave us space for tucking in the siding.

After the furring strips were nailed in, we ran the main rake boards. The level cuts where the boards fit over the return flashing were made first cuts. We then tacked the rake board in place until all the boards were cut and fit.

After working up from the corners, we made our final cut at the peak where the two rakes met. We made a plumb cut at the roof pitch on each side, leaving the length a little strong. The extra length gave us a little room to adjust the cut. Again, we used a sharp block plane to fine-tune the peak cut.

When the main rake boards were all fit and tacked in, we went back and nailed them off. The final rake-trim detail was an angled piece of molding we fabricated on site to imitate the crown on the fascia. With our peak angle all set from the main boards, we worked from the peak down, installing the final decorative trim (see bottom photo on facing page).

When the trim is installed, we do one thing to keep the painters happy. If the trim color differs from siding color, we hit the edges of the trim with finish paint to spare them the brushwork after the siding is on.

*Robert Weatherall* *builds custom homes and timber frames in Ipswich, MA.*

**Furring-strip rake spacer.** Two courses of 1x3 strapping create a space under the rake boards for the siding.
*Note:* **Scaffolding should have guardrails and toe boards to make it safer.**

## Looking up the Rake

Instead of gable overhangs, 1x3 strapping was nailed along the edge of the roof with the rake boards nailed on top. The space behind the rake boards below the strapping hides the ends of the gable siding. A strip of decorative trim milled on site finishes the rake.

Roof sheathing

Strapping spacers

Decorative rake trim

Gable rafter

Rake board

Space for siding

Gable wall plate

Wall sheathing

**Bevel-edge trim** milled on site is nailed onto the rake boards. Primer (in the bucket) is applied to the cut ends of the trim as it is installed.

# Building an Ornate Cornice

■ BY TERRY MURPHY

**F**rank Lloyd Wright didn't care for Victorian houses. He thought their abundant decoration was overdone, and their tall double-hung windows reminded him of guillotines. But I think the Victorian houses built in San Francisco during the late 19th century are among the world's most beautiful wooden structures. Newly minted, these houses bristled with layers of machine-tooled decoration, meticulously assembled by the carpenters of the Industrial Age.

Then in the mid-20th century, the city started to lose its Victorian legacy. Many of the new owners of the houses, inspired by the streamlined look of the 1940s and the 1950s (and tired of the upkeep required to maintain a Victorian house), decided to strip their homes of embellishment. They ripped off the brackets, the columns, the window hoods, and the intricate moldings to get down to the siding. The houses then were covered with a layer of asbestos shingles or a cocoon of stucco. All over the city, houses began to take on an anonymous, gray-flannel look.

This depressing trend began to reverse itself, when some visionary owners of stripped-down Victorian houses decided to pull off the stucco and restore the original detailing. These folks regarded themselves as trustees of San Francisco's architectural heritage. They weren't content to Victorianize their homes by adding a bracket here and some gingerbread there. Instead, they wanted to re-create as faithfully as possible the original look of their homes. Pam Whitehead joined the ranks of these Victorian-house liberators a couple of years ago.

## Architectural Archaeology

I first met Pam in 1987, when I was restoring a small, flat-front Victorian down the hill from her house in San Francisco's Mission District. She was impressed with the transformation she saw taking place in her neighborhood, and she started to see the potential hidden behind the stucco that cloaked her house. By 1989 she was ready to get started. I teamed up with contractor John Hornor and carpenter Bryan Kelly and started tracking down the leads that would tell us what the house originally looked like.

A good place to search for leads is the local historical society, which typically maintains an archive of old photos. Another good place to search is the neighborhood

**Restoring the cabinet grade trim** on this San Francisco Victorian home required detective skills as well as craftsmanship.

itself. Occasionally a longtime resident will have a scrapbook with photos of the surrounding houses.

We were lucky. Pam had learned that her house was the boyhood home of one of California's governors—"Sunny" Jim Rolf (Rolf had also presided over the city's rebuilding as mayor after the 1906 quake, along with some less-lofty activities). In a biography of Rolf, Pam found a small photo of his boyhood home. It showed that what had become a squat, flesh-colored mass of stucco with aluminum windows was once a restrained Victorian house of the Italianate style. When new, its octagonal bay had arched first-floor windows topped by what's called a belt cornice and a generous main cornice with a mansard roof.

I think the real character of these houses comes from their cornices—the three-dimensional projections, especially at the bays, that provide the texture and signature of the house. One of our primary challenges in restoring Pam's house was deciding what the main cornice and the belt cornice used

to look like. Figuring that out and then building them are the subjects of this article.

# Reading the Paint Scars

After the crew stripped the stucco from Pam's house, an ugly patchwork of siding, sheathing, and chopped-up pieces of the original trim boards was revealed. The main cornice was gone. The stucco crew had torn it down to make room for the hideous tile hat that crowned the house when Pam bought it.

With a photo of Sunny Jim's house in hand, I went up the scaffolding to take some measurements and to study the ghostly images of the detailing left on the original siding. Pam's house was rich with clues. The paint scars revealed the traces of arched frames around panels, a hood over the portico and the profile of the belt cornice over the first-floor windows. I photographed each piece of evidence. In some cases I pinned pieces of paper over the paint scars

*I think the real character of these houses comes from their cornices—the three-dimensional projections, especially at the bays, that provide the texture and signature of the house.*

**Tiled and defiled.** By the 1950s, many San Franciscans had stripped their houses and stuccoed them. This Italianate Victorian had been further disfigured with a barrel-tile roof.

**Sunny Jim's house.** While researching its history, the owner of this house hit pay dirt when she found its photo in a book about one of its inhabitants: California Governor James Rolf.

and rubbed a pencil on them to capture the profile of some long-lost molding.

With the tracings in my bag, I searched for pieces of the original trim. Typically the crew that stripped a house to prepare it for the stucco work would use the larger trim pieces such as window casings—nailed flat-side out—to fill in areas that needed sheathing. These pieces can be invaluable when it's time to mill the new moldings.

With the information gleaned from the tracings, the Sunny Jim photo, and the silhouettes of moldings, I prepared some drawings for the Landmarks Board. My drawings included a section of the belt cornice based on the paint scar (see drawing on facing page). For the mysterious main cornice, I relied on the Sunny Jim photo to get the proportions right. Unfortunately, some of the main-cornice details were lost in the shadows. We'd have to figure these out as we went along.

## Preparing for Trim

I matched the drawings of the paint scars with the molding catalog from San Francisco Victoriana, Inc., a company that specializes in hard-to-find Victorian trim pieces. Sure enough, I found many of the patterns we needed, so I placed an order for redwood molding stock.

With thousands of dollars worth of precious moldings soon to arrive, John and I focused on getting the siding ready for its new trim. Then we had to make a tough call. It's almost always a good idea to use the original material. However, the channel rustic siding left on this house was badly gouged and split from all the nails that had been driven into it. Despite a desire to use the original stock and the information attached to it, we couldn't bear to put all that new redwood on such a bad surface. We stripped the old skin and renailed the diagonal sheathing beneath it. Then we hung the new 1x10 channel rustic siding on the body, like a canvas waiting for paint.

## Paint Scars Reveal Original Details

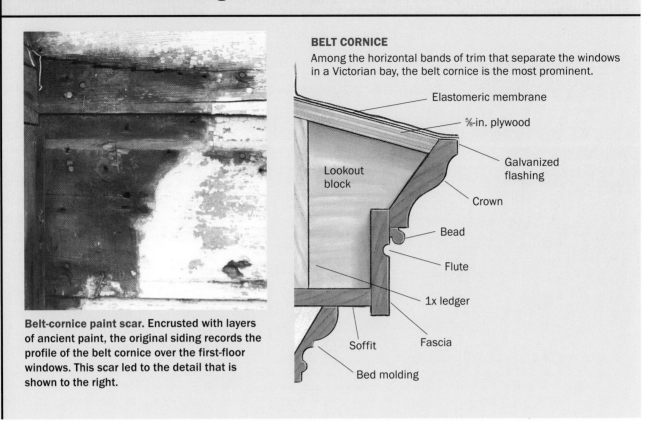

**Belt-cornice paint scar.** Encrusted with layers of ancient paint, the original siding records the profile of the belt cornice over the first-floor windows. This scar led to the detail that is shown to the right.

### BELT CORNICE
Among the horizontal bands of trim that separate the windows in a Victorian bay, the belt cornice is the most prominent.

- Elastomeric membrane
- ⅝-in. plywood
- Galvanized flashing
- Crown
- Bead
- Flute
- 1x ledger
- Fascia
- Bed molding
- Soffit
- Lookout block

## Raising the Bay

Before we could finish the siding, we had to add several feet to the height of the build-ing. But we weren't sure exactly how tall the building had to be. We decided to frame it higher than we thought necessary, then cut it back based on the Sunny Jim photo.

We began by framing an 8-ft. wall atop the existing plate above the bay window and the adjacent plates atop the straight walls. These were the plates that had been lowered when the house had been remod-eled in the 1950s. The new wall is tied to the old, sloping roof by way of a new, flat roof (see drawing on p. 20).

With the new wall in place, I took a photo-graph of the house from exactly the same spot as the Sunny Jim photo was taken. Then I figured out a scale for each photo

**Putting it back.** Guided by an old photo of the house, the author extended the parapet and the bay window to their original heights and then began work on the restored cornice.

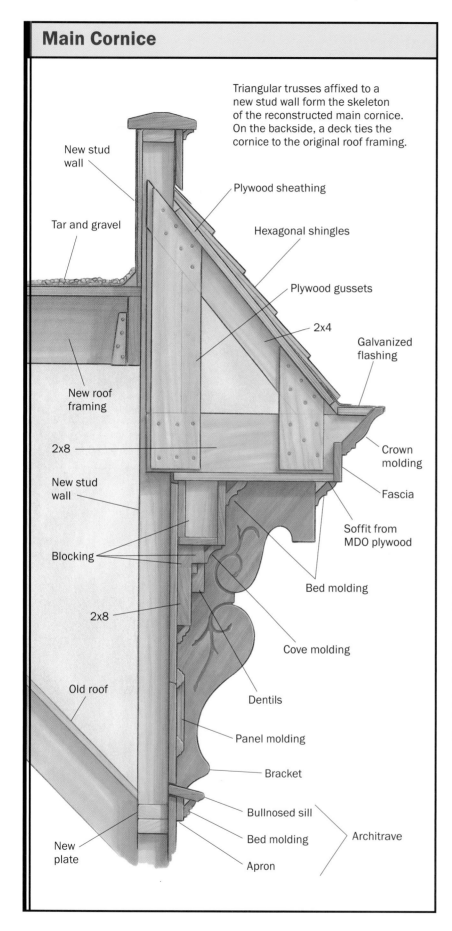

Triangular trusses affixed to a new stud wall form the skeleton of the reconstructed main cornice. On the backside, a deck ties the cornice to the original roof framing.

New stud wall

Tar and gravel

New roof framing

2x8

New stud wall

Blocking

2x8

Old roof

New plate

Plywood sheathing

Hexagonal shingles

Plywood gussets

2x4

Galvanized flashing

Crown molding

Fascia

Soffit from MDO plywood

Bed molding

Cove molding

Dentils

Panel molding

Bracket

Bullnosed sill

Bed molding

Apron

Architrave

based on the known dimensions of the details, and I drew a grid on each one. By comparing the two photos, I was able to pinpoint the top of the facade, and we lowered the new wall accordingly to that height.

Based on the proportions of the siding, I made plywood cutouts that approximated the profile of the main cornice. Then I tacked the cutouts to the framing and went back to the street to see how they looked. I fiddled with the slope and the depth of the cutouts until I had them about right. Next, I made a section drawing of the main cornice to capture the proportions and to figure out the structure (see drawing at left).

My design uses a series of small triangular trusses to support the structure of the main cornice. The trusses are made of Douglas fir 2xs held in the same plane by plywood gussets. The profile of the trusses defines the soffit, the fascia, and the mansard roof over the cornice while the beaks on the ends of the lower cords are backing for the crown molding (see photo on facing page). John and I nailed the trusses to the new stud wall and then locked them together with a soffit of medium-density overlay (MDO) plywood, a redwood fascia and a roof of ⅝-in. CDX plywood.

Affixing the trusses to the bay and then sheathing them established the shape of the main cornice—a shape that would be tough to change if it didn't look right. So as soon as John and I had tacked the sheathing in place, I clambered down the scaffolding for the moment of truth. I held up the Sunny Jim photo and compared it to the house. The shadows looked the same, and I breathed easier.

**Supporting the main cornice.** Triangular trusses made of Douglas-fir 2xs fan out from the wall framing to serve as backing for the cornice roof and soffit. The trusses are held in the same plane by plywood gussets. The beaks on the bottom cords of the trusses anchor the crown molding.

## Installing the Belt Cornice

The belt cornice over Pam's first-floor windows is a typical detail of the Victorian era. It is composed of a fascia, a crown, and a soffit, all of which are topped with a narrow roof. Installation began with the lookout blocks, which provided nailing surface.

We used a template to lay out the lookout blocks, which we cut out of rough-sawn 1x redwood. Then we affixed the blocks on 1-ft. centers to 1x ledgers. Nailing the blocks to the ledgers before hanging them on the wall makes it easy to get the blocks aligned. Because they're often going on buildings that are sagging from age, I can't always install the blocks level. But I sure try to keep things plumb, square, and level if possible. To do otherwise starts a string of costly out-of-square trim carpentry that will require shims, asymmetrical miters, and scribed cuts.

The belt cornice went together starting with the soffit. The fascia was applied next. It ends ⅜ in. below the soffit to create a drip edge. Next came the crown, which is the most fun because it's the most dramatic part of the assembly. Where these pieces met at the miters, they were cross-nailed and glued with yellow glue. To finish the crown, I nailed on the beading to approximate the profile of the crown in the paint scar.

The ⅝-in. plywood lid came next. We beveled its leading edge to match the angle formed by the back of the crown. This narrow lid is covered with galvanized flashing and a strip of elastomeric membrane called Gacoflex, which tucks under the counter-flashing above. The last element in the belt cornice is the bed molding, which flares out from the wall to engage the cornice's soffit.

## Making the Brackets

A dozen big brackets and seven small ones, called modillions, march across the main cornice, appearing as if they held up the mansard roof (see top photo on p. 23). These brackets have five layers—three 1xs sandwiching two 2xs—for a total thickness of 5¼ in. The 2xs are recessed along most of their exposed edges, creating a shadow that emphasizes their curves. I built the brackets and modillions, including the fretwork incisions in the outer layers, on site with templates and a router.

## Sources

**Gaco Western, Inc.**
**(800) 456-4226**
www.gaco.com
*Gacoflex*

**San Francisco Victoriana, Inc.**
**(415) 648-0313**
www.sfvictoriana.com

I began by spray-gluing a full-size drawing of the bracket to a piece of ½-in. plywood. Cut out with a jigsaw, this pattern became the template that I used to guide my router as I cut out the bracket pieces (see bottom left photo on facing page). I have a long-shanked ½-in. bit that plows through redwood in a hurry, so I used it to cut out the bracket profiles. Then I switched to a ¼-in. bit and collar (see bottom right photo on facing page) to rout the fretwork pattern in the exposed side of the brackets.

To save on materials, the 12 big brackets are hollow. After sanding the rough edges, I assembled the bracket parts from the inside out with glue, nails and screws. Then I set them aside until we were ready for them on the high wall. I made the modillions, which fit between the brackets, using exactly the same tools and materials.

## Main Cornice

John and I knew trimming the main cornice would take a couple weeks, so we began by making a place to work. John placed some scaffolding planks on the top tier and built a workbench that was large enough for our sliding compound-miter saw. If you've ever scrambled around on narrow scaffolds 30 ft. above the sidewalk, you can appreciate the importance of a generous workspace on the top step.

First we assembled the architrave—the lowest part of the cornice—from a bullnose sill, a 1x6 apron, and a bed molding. The tails of the brackets bear on the top of the architrave (see top photo on facing page). We fastened the brackets plumb and square under the soffit, using a pneumatic nailer to position them and angled screws to anchor them. A box made of 1x6s tucks under the soffit, running between the brackets and bracing their sides. The box appears to run continuously across the front of the house, filling out the cornice and providing anchorage for the modillions.

After extending the 1x6 box across the cornice and placing the modillions, John began to cope and miter the dozens of pieces of 2-in. bed molding that wrap the perimeters of these features. Meanwhile, I attached a 2x8 ground for my favorite feature in the cornice, the dentils.

The dentil is an antiquarian motif resembling a string of tiny rectangles. Perhaps its most famous application occurred at the Acropolis. Typically paired with a bed molding above, dentils appear to be a series of alternating small and large blocks. In reality, the Victorian builder created the dentils by spacing the notched larger block over a continuous smaller strip. We used the same method. We ran a ¾-in. by 1¼-in. strip of redwood under the bed molding. Over this strip we nailed notched blocks cut from ripped and rabbeted 2x6 stock.

As I figured out the spacing of the dentils on my solar-powered calculator, I couldn't help but muse over the irony of laying out trim from antiquity with solid-circuit technology. But it works.

The dentils turned out to be just the right touch for the smaller-scale layer of detailing in the cornice. They may very well have been the detail hidden in the shadows of the Sunny Jim photo. We nailed on the panel moldings to frame the frieze areas between the brackets, which completed the main cornice (see photo on p. 17).

*Terry Murphy* is a designer/builder specializing in Victorian restoration.

**Cabinet-grade trim at 30 ft.** Curvy brackets and modillions (smaller brackets), which are laminated from alternating layers of 1x and 2x redwood, detail the cornice's soffit and frieze. The dentils provide a secondary level of texture. On top of the architrave, two dentils await installation.

**Making the brackets.** Over 100 pieces went into the brackets and modillions, making their fabrication perfect for template routing. In the photo at left, the author routs the profile of an outer lamination. The next pass, with a ¼-in. bit and collet (above) will mill the fretwork pattern in its face.

# Exterior-Trim Details That Last

■ BY JOHN MICHAEL DAVIS

**Drainage mat gives moisture a way out.** Because even the best caulk joint can fail, a ⅜-in.-thick drainage mat is applied betweeen this porch post and its base trim. Any moisture that gets past the caulk is able to run out beneath the front baseboard, which is left uncaulked.

The old saying "It's better to be lucky than smart" certainly applies to New Orleans. Considering the rot-acceleration chamber that passes for a climate here, it's remarkable that New Orleans has more 19th-century houses than any other city in the country. Although it's tempting to credit the skill of old-time carpenters who built things to last, the only reason all these houses are still standing is luck. When the first Europeans arrived, almost every acre of land in New Orleans was concealed beneath a massive canopy of old-growth cypress trees. Clearing the land for development released a seemingly unlimited supply of one of the heartiest building materials on the planet.

The old-timers didn't worry about back-priming, drainage planes, or caulking. They just nailed together two pieces of wood and walked away; no fungus was going to take a bite out of a 2,000-year-old chunk of cypress. Do that with the fast-growth sapwood we have to work with today, and you can be sure the forces of decay will start moving in before the extension cords are rolled up.

## Caulk Is the First Line of Defense

Among all the places where rot can secure a foothold in exterior woodwork, caulked joints are probably the most vulnerable. Seasonal changes in temperature and humidity cause wooden trim elements to expand and contract. Unless caulk has the flexibility to accommodate this movement, sooner or later, the bead will crack, and water will seep in. From that point on, the caulk actually does more harm than good, because it allows water to soak into the wood while it restricts air circulation that would promote drying.

My approach to protecting vulnerable exterior trim is twofold: First, I try to create a caulk joint that can weather many seasons of expansion and contraction. I've been around long enough to know that even the best caulk joint will eventually fail, however; so as often as possible, I also build a path for moisture to escape when that happens.

## Backer Rod Gives Caulk the Freedom to Move

There are two ways to create a caulk joint that can handle expansion and contraction: Either reduce the amount of movement or increase the ability of caulk to move. I do both. Certain species of wood, such as pressure-treated southern yellow pine, are inherently unstable, so I try not to use them when given a choice. For exterior trim, I prefer to use the highest-quality kiln-dried

*Even the best caulk joint will eventually fail; so as often as possible, I build a path for moisture to escape.*

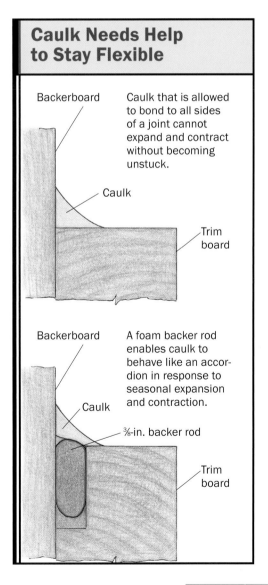

### Caulk Needs Help to Stay Flexible

Backerboard

Caulk that is allowed to bond to all sides of a joint cannot expand and contract without becoming unstuck.

Caulk

Trim board

Backerboard

A foam backer rod enables caulk to behave like an accordion in response to seasonal expansion and contraction.

Caulk

⅜-in. backer rod

Trim board

lumber I can find; in my area, that's either mahogany, Spanish cedar, or all-heart redwood. I also fully prime every piece of trim, and I assemble permanent (crackproof) miter joints using biscuits and marine epoxy.

I increase the ability of the caulk to move with the wood by applying high-quality polyurethane caulk (see sidebar on p. 33) on top of a backer rod to create a two-sided caulk joint. (Caulk doesn't adhere to the foam backer rod.) The biggest mistake most amateur caulkers make is to fill the joint completely where two pieces of wood meet. This procedure not only wastes caulk but also creates a three-sided joint that's prone to failure: The caulk adhering to the bottom as well as to both sides of the joint is left little room for movement. A two-sided joint—one in which the caulk bonds to the sides but not the bottom of the joint—allows the caulk to expand and contract like an accordion.

In principle, creating a two-sided caulk joint is simple enough: Just cover the bottom of the joint with a bond breaker, any material that will prevent the caulk from bonding. If a joint is shallow and cannot be enlarged, special Fine Line tapes are used as bond breakers, but these tapes can be difficult to work with. An easier, better approach is to enlarge the joint to make space for a foam backer rod.

Wherever there's a place that two pieces of exterior trim must fit together tightly, my standard operating procedure is to cut a ³⁄₁₆-in.-wide by ½-in.-deep rabbet into the inside edges of the trim pieces (see drawing on p. 25). After the trim is assembled, I compress a ⅜-in.-dia. poly foam backer rod into the bottom of the joint. This ⅜-in. backer rod leaves enough space on top for a ³⁄₁₆-in.-wide by ⅛-in.-deep bead of caulk.

## Erosion Mat Gives Moisture an Out

Unfortunately, the best caulking job doesn't come with a lifetime guarantee. No matter how conscientious I am, at some point in time, somewhere along the joint, I know the caulk is going to fail. Failure could be caused by a minor installation flaw such as a drop of sweat on the wood or an undetected bubble in the bead. If I'm lucky, the joint will last the life of the sealant (20 years tops). But at that point, no one else is going to lavish the same attention on the work as I did. So to ensure long-term survival of exterior trim dependent on caulk for survival, I build in a means for moisture to escape.

An escape route for moisture could be something as simple as a couple of circular louvers near the top and bottom of a hollow column to promote air circulation. On the other hand, a fully exposed trim element such as a corner board, drip cap, or porch skirting (see drawing on facing page) that has nothing but caulk to protect its innards from wind-driven rain or splash-back requires an active drainage plane.

Not long ago, I had to repair built-up posts that supported a railing on an uncovered second-story porch. In this case, most of the rot damage occurred when water seepage became trapped between the applied baseboards and the bottom of the post carcase. To prevent the same type of damage from happening in the future, I created a simple drainage plane to give moisture a way out (see drawing on p. 28).

Any type of rot-resistant shim stock can be used to create a drainage plane, but the best material I've found is Enkamat #7010, a ¼-in.-thick mat of entangled nylon filaments that is manufactured for use as an erosion matting (see sidebar on p. 32). After cutting the mat to size with a utility knife, I wrapped it tightly around the post and then fastened it with staples.

## Active Drainage Planes Protect Vulnerable Trim

These drawings depict just three examples of how well-planned caulk details used in combination with an active drainage channel offer two layers of protection to the most-exposed parts of a house.

### CORNER BOARDS (PLAN VIEW)

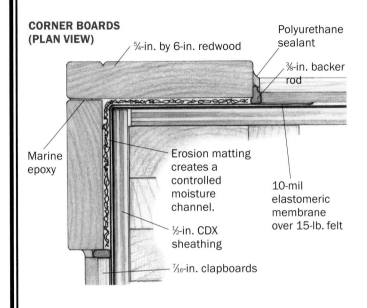

5/4-in. by 6-in. redwood

Polyurethane sealant

3/8-in. backer rod

Marine epoxy

Erosion matting creates a controlled moisture channel.

1/2-in. CDX sheathing

10-mil elastomeric membrane over 15-lb. felt

7/16-in. clapboards

### PORCH SKIRTING (ELEVATION)

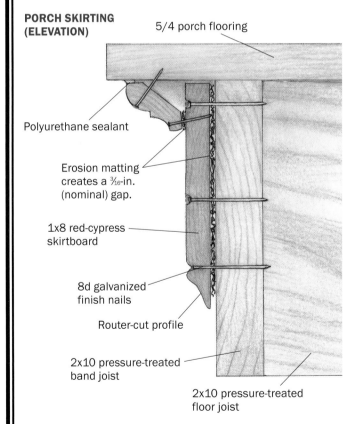

5/4 porch flooring

Polyurethane sealant

Erosion matting creates a 3/16-in. (nominal) gap.

1x8 red-cypress skirtboard

8d galvanized finish nails

Router-cut profile

2x10 pressure-treated band joist

2x10 pressure-treated floor joist

### WATER TABLE AND DRIP CAPS

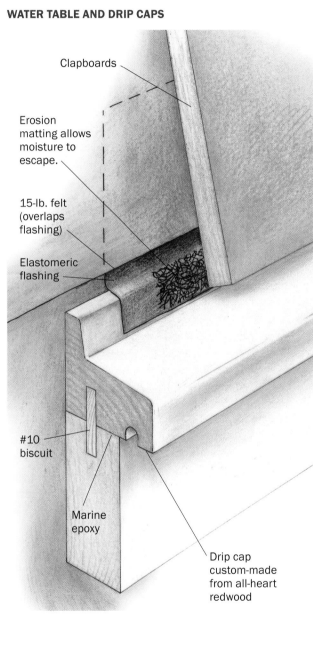

Clapboards

Erosion matting allows moisture to escape.

15-lb. felt (overlaps flashing)

Elastomeric flashing

#10 biscuit

Marine epoxy

Drip cap custom-made from all-heart redwood

# Two Ways to Let Moisture Escape

1. Leaving the front baseboard uncaulked at the bottom promotes drainage and air circulation.
2. The angle brackets that secure the newel cap to the post create a ¹⁄₁₆-in. gap that allows humid air to escape.

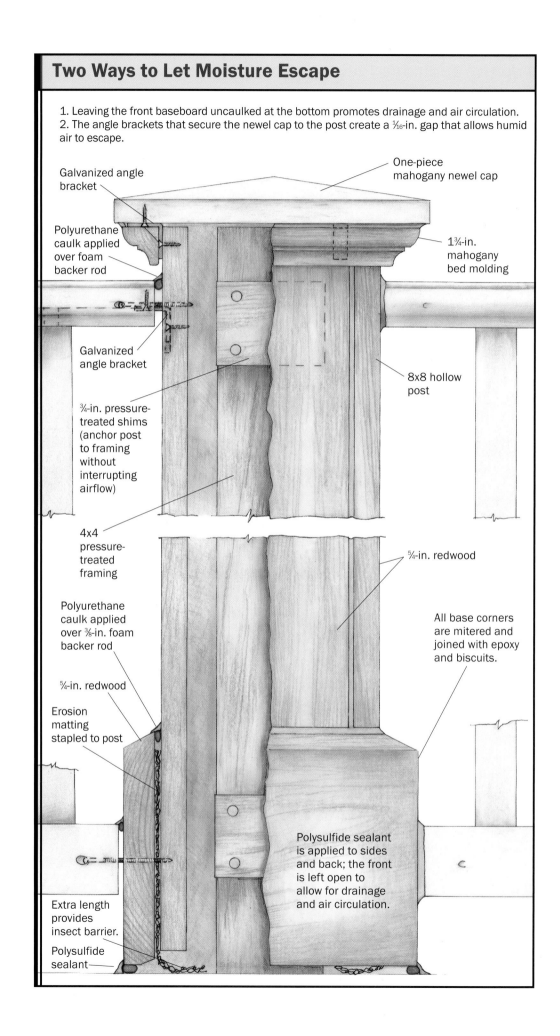

Galvanized angle bracket

One-piece mahogany newel cap

Polyurethane caulk applied over foam backer rod

1¾-in. mahogany bed molding

Galvanized angle bracket

¾-in. pressure-treated shims (anchor post to framing without interrupting airflow)

8x8 hollow post

4x4 pressure-treated framing

⅝-in. redwood

Polyurethane caulk applied over ⅜-in. foam backer rod

All base corners are mitered and joined with epoxy and biscuits.

⅝-in. redwood

Erosion matting stapled to post

Polysulfide sealant is applied to sides and back; the front is left open to allow for drainage and air circulation.

Extra length provides insect barrier.

Polysulfide sealant

To allow for the mat's thickness, I added ⅜ in. to the length of each baseboard. After cutting all the baseboards to length, I pre-assembled three sides of the box before slipping it around the post. While applying slight pressure on the corners to compress the mat, I checked the reveals by sight before I tacked the baseboards using an air nailer (see photo on p. 24).

After the final side of the baseboard was glued and tacked in place, I filled the gap around the top with ⅜-in. backer rod. The best tool I've found for forcing backer rod into a tight gap is a window-screen spline roller (see photo 3 on p. 30). Because this tool has different-size wheels on each end, I choose the end that fits the situation best and simply roll the rope in with one hand while stretching it slightly with my other hand. Using a finger as a gauge, I roll back and forth over the backer rod until the top is at least ⅛ in. below the top of the baseboard. Then I cover the backer rod with a ³⁄₁₆-in. bead of polyurethane caulk ( see photo 4 on p. 30).

# Hardware Creates Breathing Room

Other weak points on the built-up post are the post-cap assembly and the joint where the railings meet the post. If the railing I'm installing is thick enough, I'll make space for ⅜-in. backer rod by rabbeting the edge the same way I described earlier. If I'm installing a narrow railing, like the one on this job, however, I cut the railing ¼ in. short and mount it on top of a galvanized angle bracket (see drawing on facing page).

Those handy angle brackets also make it easy to create a breathing space while mounting the post cap. Using a scrap of cardboard as a temporary shim between the top of the post and the cap, I installed two 1-in. angle brackets on each side to anchor the post to the cap; then I removed the shim

and covered the brackets with a bed molding. I caulked the joint where the molding met the cap but left the bottom of the molding uncaulked to promote air circulation.

# Older Trim Needs Help, Too

Although modern lumber is more susceptible to rot, even old-growth cypress can deteriorate after years of neglect. When I have to repair localized rot damage on significant structures such as windowsills or porch columns, I try to rebuild the missing or damaged sections in place using epoxy. Wherever a failed caulk joint was the source of moisture, I retrofit the structure with working caulk joints. I also try to install air vents and weep holes that could help to promote drying (see drawing on p. 31).

Recently, I used these strategies to create working caulk joints on the base of a 10-in. round column. As with most cases, the worst rot damage occurred at the top and the bottom of the base. At these places, I used a couple of specialty tools (not made for the purpose) to carve out the space for backer rods.

At the top of the base, where the column shaft rested on the upper torus ring of the base, I was able to use a corner-grooving tool, which is intended for installing weather-stripping in old door frames, to carve a ⅛-in.-wide by ¼-in.-deep dado (see top left photo on p. 31). I've also been able to make this cut, although not as easily, using a cordless drill equipped with a ⅛-in. ball-shaped die-grinder bit. The only problem with carving a narrow groove such as this one is that standard backer rods are too wide; fortunately, I have found an alternative, which I'll describe in a moment.

## Sources

**Abatron**
(800) 445-1754
*PrimKote #8006-1*

**Charette**
(800) 367-3729
*Palette knife*

**Colbond Geosynthetics**
(800) 365-7391
*Enkamat #7010*

**Crain Power Tools**
(408) 946-6100
*Flush cutting saw*

**Gougeon Bros., Inc.**
(989) 684-7286
*Marine epoxy*

**Midget Louver Co.**
(800) 643-4381
*Midget Louvers and thimble vents*

**Resource Conservation Technology, Inc.**
(410) 366-1146
*Sikaflex 1a, corner grooving tool*

**York Manufacturing, Inc.**
(207) 324-1300
*Wascoseal*

# Creating the Perfect Caulk Joint

Before I apply any caulk, I make sure all the wood surfaces are fully primed; if some of the woodwork in the photos looks unprimed, it's because my favorite primer (Primkote #8006-1) is a two-part epoxy that goes on clear. Wherever I've installed backer rod, I use my finger as a gauge to make sure the rods are deep enough to allow a minimum sealant depth of ⅛ in.

A professional-grade caulking gun is a must for applying thick polyurethane sealants (photo 4); I've never given much thought to whether it's better to push or pull the gun while applying the caulk, however, because to me the gun is simply a delivery device. I tool every bead of sealant I apply, and my favorite tools are my thumb and forefinger. (They're just always there; what can I say?)

Polyurethane has to be cleaned up with mineral spirits, and over the years, I've developed chemical sensitivity to prolonged exposure, so I wear 4-mil disposable nitrile gloves almost all the time. It's not always easy to get crisp drags with gloves on, so when the appearance of the joint is crucial, I will use a bare finger if nothing else works, but I prefer to use a tool.

I think I've tried every tool specifically designed for dragging caulk, and I've never had much luck with any of them. What works best for me is a 3-in. artist's palette knife (photo 1). Looking like a miniature bricklayer's trowel, a palette knife has a long, thin, tapered blade with a rounded tip that's flexible enough to offer precise control. The blade is also polished and slick enough to get a really smooth drag, especially if it's regularly wiped off and lubricated with solvent.

Palette knife tools visible joints.

**1**

Mineral spirits remove excess caulk.

**2**

Backer rod allows caulk to flex.

**3**

Polyurethane is the first line of defense.

**4**

# Modern Materials Protect Historic Trim

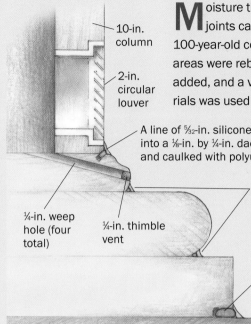

- 10-in. column
- 2-in. circular louver
- A line of ⁵⁄₃₂-in. silicone tubing is inserted into a ⅛-in. by ¼-in. dado (cut with router) and caulked with polyurethane sealant.
- ¼-in. weep hole (four total)
- ¼-in. thimble vent
- A piece of ⅛-in. bond-breaking tape lines the bottom of the gap and is caulked with polyurethane.
- A piece of ⅜-in. foam backer rod is inserted into a ¼-in. by ½-in. dado (cut with flush-cutting trim saw) and caulked with polyurethane.
- Brick pedestal

**M**oisture that seeped in through failed caulk joints caused extensive rot damage to this 100-year-old column base. After the damaged areas were rebuilt with epoxy, air vents were added, and a variety of modern tools and materials was used to retrofit working caulk joints.

**A corner-grooving tool** carves a ⅛-in. dado into the joint where the column shaft meets the upper torus of the base.

**Custom-mounted on a Plexiglas base,** a flush-cutting saw cuts a rabbet where the edge of the plinth meets the brick pier.

**When standard backer rods are too thick,** other materials are pressed into service. Some ⁵⁄₃₂-in. silicone weatherstripping fits the ⅛-in. groove at the top of the base.

**Enkamat #7010 erosion matting** creates a drainage channel between trim boards.

# The Right Stuff May Be Hard to Find

**A**lthough some of the materials mentioned in this article can be found at local supply houses (depending on your location), many are available only by mail order. So plan ahead.

**Foam (⅜ in.) is the standard backer rod for two-sided caulk joints.**

**Silicone tubing (⁵⁄₃₂ in.) serves as alternate backer rod for narrow (retrofitted) caulk joints.**

**Enkamat # 7010 erosion matting creates a drainage channel between trim boards.**

**Wascoseal® 10-mil elastomeric membrane serves as flashing behind drainage.**

**Midget Louvers (1 in.) and thimble vents (¼-in.) allow moisture to escape from within hollow columns.**

**Fine Line tape (⅛ in.) serves as bond breaker for shallow caulk joints.**

Before completing the caulking detail on the top of the column base, I used the flush-cutting saw I'd normally use for trimming door jambs to make a space for standard backer rod underneath the base (see top right photo on p. 31). The flush-cutting saw allowed me to place a ½-in.-deep saw kerf, ¼ in. up from the bottom. I needed a few swipes with a sharp chisel to remove the waste, and the base was ready for backer rod.

After blowing out the sawdust, I coated both grooves with Primkote #8006-1, a fast-drying epoxy primer that did not clog the groove as most primers would have. Five minutes later, when the primer was dry, I inserted a ⅜-in. foam backer rod in the groove beneath the base. In the narrow groove on top, I inserted my stand-in for backer rod, a length of 5⁄32-in. silicone tubing (see bottom left photo on p. 31) that, like the cutting

tool, was also intended for weatherstripping (see sidebar on the facing page).

Although the top and bottom of this column suffered the most rot damage, I also caulked the two middle joints where the individual torus sections came together. Both joints had enough cleavage between them for caulk but not backer rod, so I used bond-breaking tape instead (see bottom right photo on p. 31). To prevent a three-sided bond that would crack and fail, I lined the bottom of each joint with ⅛-in.-wide (#218) Fine Line tape. Once all the bond breakers were in place, I caulked all the joints with polyurethane sealant and tooled them smooth, confident that this column will be around a lot longer than I will.

*John Michael Davis* is a restoration carpenter in New Orleans LA

## The Best Caulks Are Flexible and Paintable

Polyurethane sealant is my preferred exterior caulk. Although more than twice as expensive as premium acrylic latex, polyurethane's vastly superior adhesion and flexibility make it a bargain. I use Sikaflex® 1a, a European polyurethane available only through marine suppliers or by mail order. Recently, my local suppliers have begun stocking polyurethane sealants from Macklanburg-Duncan. One day, I'll try them.

Although polyurethane is my standard exterior caulk, for super-critical joints—such as the front edge of a threshold—I use polysulfide. Polysulfide costs three times as much as Sikaflex and takes 3 days to 7 days to dry. Nevertheless, it retains significantly more flexibility far longer than polyurethane and is sandable.

Where appearance is critical, sandability allows me to create seamless fillets. I use 3M® #101 polysulfide but buy (3 oz.) tubes because they don't keep well after opening.

Silicone is highly flexible and adhesive, but I don't use it often because—despite some marketing claims—it can't be painted. Where wood has been painted or where trim elements are metal, glass, or masonry, silicone is a more effective caulk than polyurethane. But all silicones are not alike. In my opinion, Dow Corning® #795 is thicker, more adhesive, and more toolable than anything I can get at a hardware store, and it costs only a dollar or two more a tube.

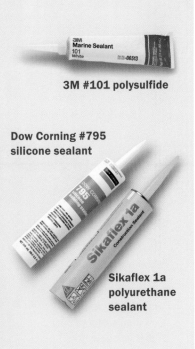

**3M #101 polysulfide**

**Dow Corning #795 silicone sealant**

**Sikaflex 1a polyurethane sealant**

# Synthetic Stucco

■ BY STEVE CULPEPPER

**What went up must come down.** Although this EIFS-sided home in Wilmington, North Carolina, suffered only minor damage from water intrusion, the synthetic-stucco skin is being ripped off anyway—to make way for more traditional siding.

What began a few years ago as a frenzy of business for installers of synthetic stucco in North Carolina has spawned a new trade: synthetic-stucco removal.

Synthetic stucco, or EIFS (exterior insulation and finish systems; the acronym is pronounced *eefs*), is being ripped from houses in almost every neighborhood in Wilmington, North Carolina. In its place: clapboards, shingles, hardboard siding, real stucco, or brick. Rarely is EIFS going back up. The siding is being replaced because water got behind the synthetic stucco and damaged the sheathing. In some cases, even the framing suffered damage.

I found this panicky removal of EIFS at odds with much of what I knew about the material. I began researching an article on synthetic stucco a couple of years ago with the goal of writing about the material, how it's applied and detailed and how it's maintained. Since then, the problems in North Carolina surfaced, and the story was no longer quite so cut and dried. From what I've since learned about EIFS, I'm not so sure I'd want it on my house—at least not the way it's been detailed in the past.

As a modern building material, synthetic stucco has a relatively long history. It was first used in post–World War II Germany to resurface buildings with damaged masonry. It came to this country in the late 1960s, and its use became widespread. In 1995, installers applied more than 260 million sq. ft. of the material in the United States, about 80% of that on residences. If applied properly, EIFS can be an attractive, low-maintenance siding. If not, it can be a disaster.

What began as a sticky issue local to North Carolina has blossomed into a small but thorny national one. On one side are EIFS manufacturers and the EIFS Industry Members Association (EIMA); on the other side are builders and contractors burned by their association with EIFS. In the middle are property owners. Of course, lawyers are present in the EIFS controversy, too, feeling their way through an evolving legal crisis that has spawned many individual lawsuits and a half-dozen class-action suits representing numbers of homeowners.

## Why Does Water Cause So Much Damage When It Gets Behind EIFS?

Traditionally, residential EIFS refers to a surface-sealed, water-barrier siding system made up of a layer of foam-insulation board, fiberglass mesh, one or two base coats, and a finish coat of synthetic stucco (see photo below).

Commercial EIFS is generally thicker than residential EIFS and usually is applied over steel framing. For residential applications, the insulation board usually is glued onto the substrate, which can be plywood, gypsum, oriented strand board (OSB), cement board, or masonry. Next comes a layer of

*The usual suspects in water leaks are to blame for water intrusion in EIFS: windows, doors, flashings— anywhere the siding is penetrated.*

**These layers should add up to a protective barrier against the elements. Glued to the sheathing is the expanded polystyrene insulation board. Over that is a base coat embedded with protective fiberglass mesh. Topping off the system is a textured acrylic finish.**

fiberglass mesh, which is bedded over the insulation with an acrylic portland-cement base coat. A second base coat sometimes is applied before the finish coat goes on. Standard EIFS also relies heavily on proper detailing, especially caulking and flashing.

At $4 to $8 per square foot, EIFS is marketed as being price competitive with brick and other masonry sidings. It expands and contracts with the seasons, its color goes all the way through the textured finish, it insulates, and it serves as a barrier to keep out the weather. Details can come in as many shapes as polystyrene can be cut into (see top photo below). Above all, EIFS looks like stucco but resists cracking, and it is supposed to be a low-maintenance siding.

Ironically, the greatest strength of EIFS is also its greatest weakness. Because it is a water-barrier system, water can't get in—

theoretically. When water does get in, it cannot easily get out. And when water gets trapped between sheathing and an otherwise impermeable covering, rot can occur (see bottom photo below).

Water intrusion is not unique to EIFS. Water can get inside almost any type of building envelope, and usually does. What's really important is that the water has a way to get out. Take cedar clapboards. The sheathing beneath clapboards is covered with building paper or housewrap; clapboards are nailed over that. Any water that blows around or through the clapboards hits the paper and dribbles out.

Standard cement stucco is applied over a substrate wrapped with waterproof building paper, to which is affixed metal lath. Three coats of stucco—scratch, brown, and finish coats—are applied over the lath, which provides a gap between the building paper and stucco for water to drain out through a weep screen installed at the foundation.

In most residential EIF systems, there is no building paper or housewrap under the synthetic stucco because the insulation is glued directly to the sheathing. Because typical flashing isn't used with most residential EIFS, caulk becomes the first line of defense at penetrations. If water gets between the sheathing and the insulation board, it can stay there for a long time, travel along gluelines, and cause damage far from the leak. In contrast, wood siding generally is caulked at doors and windows, where caulk serves as a secondary barrier against moisture. With wood siding, the main barrier against moisture is the building paper–covered sheathing.

The usual suspects in water leaks are to blame for water intrusion in EIFS: windows, doors, flashings—anywhere the siding is penetrated.

**Using a template and hot-wire cutter, an EIFS installer can slice through a thick piece of polystyrene to make almost any type of molding or detail.**

**This Cary, North Carolina, house was about 1 year old when moisture-testing showed that water had gotten behind the synthetic stucco. Inadequate seals around windows caused most of the leakage. Water dribbled past the caulking or through the jamb-and-sill connection to the OSB substrate.**

# Water Can Get in Where EIFS Terminates

Unfortunately, the highly publicized failures in North Carolina call into question the materials and installation details of all EIFS. In Wilmington, the local American Institute of Architects (AIA) surveyed 209 EIFS-clad houses—most just a few years old—and found 90% with moisture-content levels above the acceptable limit of 19%. (Moisture levels below 19% require no further action, according to a National Association of Home Builders, EIFS report).

Facts in the AIA report damaged the EIFS industry argument that the problem was caused by a few crummy builders and installers. Houses surveyed by the AIA were in 16 different subdivisions, built by 19 different builders, had EIFS applied by 10 different applicators, and used 12 different EIF systems.

However, EIMA's charge of poor detailing in Wilmington was at least partly on target. Of houses sampled by the AIA, 38% had improper caulking or no caulking around windows. That lack of proper caulking might not have been a problem had the houses been sided with a system that allowed water to drain out.

New Hanover County—where Wilmington is located—stopped accepting EIFS as an alternate to code-recognized building methods on February 15, 1996. For now, the county allows EIFS only when there is a way for water to drain out. On May 1, the state capital city of Raleigh issued a notice to builders that its inspections department would no longer accept EIFS applied over wood-frame buildings as an alternative material or method. Beginning in 1997, rules for EIFS throughout North Carolina require synthetic stucco installed over wood-frame construction to include elements to allow water drainage.

Code organizations and the American Society of Testing Engineers (www.astm.org) have developed detailed guidelines for installing EIFS.

# In Some Places, EIFS Works as Planned

In Spokane, Washington, synthetic stucco seems to have worked well. Architect/builder Gerry Copeland has put EIFS siding on more than a dozen houses in and around Spokane over the past decade. He's never had a callback.

Except for a few dings and some cracks, the EIFS cladding that Copeland's subcontractors applied (and that he and I probed with a moisture meter) appeared to have held up well. However, experts say it's easy to miss areas with high moisture levels, so we could have overlooked areas of wet sheathing.

Clearly, synthetic stucco can work, or more precisely, it can be made to work, at least in some climates. In Spokane, I spent a few days with Bob Bogart, who owns Wall Tec Systems. Bogart has been applying Dryvit's® EIFS for over ten years and has done most of Copeland's EIFS applications.

I'll outline the steps Bogart followed when installing EIFS on a house in Spokane.

**Architectural elements** such as this EIFS-clad tower can be made to look like real cement stucco.

First, there are a few things you need to know. Before the insulation board is attached to the substrate, the substrate should be flat, well-secured, and free of defects. Next, a strip of fiberglass mesh is applied to the sheathing wherever the insulation board will terminate: at the foundation, at the roofline, and at penetrations. Once the insulation board is attached, the mesh will be wrapped at least 2½ in. back around the edge of the insulation to protect it. This process is called back-wrapping.

Most manufacturers make mesh in a variety of strengths, generally from 5 oz. to 20 oz. per square yard. Ordinary mesh works in most applications, but around doors and high-traffic areas, a heavier-duty mesh should be used.

For most EIFS installations, the insulation is ¾-in.-thick polystyrene, which also is used in thicknesses up to 4 in. Bogart uses polyisocyanurate board. The material costs

more but is a lot more durable and resists denting better than polystyrene, he said.

Bogart mechanically fastens the insulation board to the substrate using Wind-Lock fasteners, which are equipped with plastic flanges. Most residential EIFS is glued to the substrate. However, in high-wind areas, insulation board may be both glued and mechanically fastened to the substrate.

Once the insulation is secured—in a staggered, running-bond pattern—Bogart and his crew head around the house, plastering on the cementitious base coat, which serves as the system's main water barrier. The base coat must be a minimum thickness of $\frac{1}{16}$ in. (Sto® Corp. recommends a ⅛-in.-thick base coat.)

## Embed Fiberglass Mesh Fully in the Base Coat

The EIMA recommends that fiberglass-reinforcing mesh be embedded in the base coat until no mesh color is visible (see photo at left). Some researchers say that after you wait at least 24 hours, a second base coat should be applied until no mesh pattern can be seen.

Bogart adds impact protection around doors and windows by attaching small, diagonal strips of mesh, or butterflies, at corners of windows and doors and around all other penetrations before applying the mesh. In cases where the EIF system passes below the floor level or on large runs of wall, a contraction joint is needed. It is caulked like an ordinary contraction joint.

The base coat must then dry for at least 24 hours. Dryvit, the largest manufacturer of EIFS in the United States, recommends the temperature be above 40°F and rising at the time the base coat and textured finish are installed (45°F and rising for nontextured finishes). This temperature must hold for at least 24 hours after installation. The material should also be protected from the elements

**A thin layer of base coat** is applied to the insulation before the fiberglass-reinforcing mesh is rolled onto the wall surface. More base coat is applied over the mesh.

for at least 24 hours. In other words, it's a lot like applying paint.

The industry recommends that during installation, EIFS be protected from dust, dirt, moisture, freezing, and humidity. Tops of walls should be covered immediately with cap flashing and final trim or else temporarily protected.

## Before the Finish Coat Comes the Caulking

In most cases, caulking means the difference between a successful EIF system and one that lets in water. It used to be recommended that caulk be applied to the finish coat at all terminations. But the finish coat proved to be an inadequate surface for caulking adhesion. Now it's recommended that caulk be applied to the base coat (see top photo).

Caulk joints should be no less than ½ in. wide and should adhere only to the insulation board and the adjoining surface, never to a third surface, such as the substrate. In cases where the depth of the joint is too great, backer rods (thin strips of foam) should be used to fill the joint.

To fill joints between EIFS and other elements, Bogart sometimes uses Willseal, an expanding insulation tape. The tape is made to be used by itself without caulking, although the manufacturer says caulking can be used. The tape is kept cold until it's installed; as it warms, it expands to a depth of about 1 in.

## Finally, a Finish Coat Is Troweled On

On a small job such as the one shown here, Bogart usually assigns one person to apply and texture the finish coat (see bottom photo). That way, the texture that's troweled into the mix will be consistent throughout. On a larger job, one person applies the finish-coat mix and another trowels in the texture.

**Caulk the base coat, not the finish.** Sealant recommended by the manufacturer is applied wherever the EIFS intersects other materials, such as at this strip of molding separating the EIFS from the clapboard siding.

**To ensure a uniform appearance,** the same person should trowel the final coat of stucco for the entire job. On big projects, one worker can apply the finish while a second worker textures it.

Weather plays a similar role in finish-coat application. The weather can't be too cold or wet, and the finish coat should not be applied in direct sunlight, which can cause uneven drying and hairline cracks. Also, it may be necessary to protect the wall surface with tents. If the temperature is below 40°F, heaters can be used inside tarps to allow application of the finish coat.

The finish coat ordered by the applicator orders depends on the texture specified. The most common textures are a smooth finish,

*In most cases, caulking means the difference between a successful EIF system and one that lets in water.*

a sand finish, a rilled finish, and a troweled finish. Also like paint, EIFS comes in a wide range of silly-named colors, such as Wild Cattails, Smoked Putty, Good Earth, and Broadway Blue.

## The Devil Is in the Detailing

EIFS manufacturers seem to see their products only in terms of their insulation board, mesh, base coat, and designer-color top coat. They don't seem to see their EIFS components in a context of windows and doors and all other components that, along with synthetic stucco, make up the exterior of a house. I asked a representative of one of the largest EIFS manufacturers about detailing around windows. She suggested I talk to a window manufacturer.

For their part, builders need to understand that an EIF system isn't like ordinary siding. Clapboards, brick, and real stucco all withstand water. Inevitably, when water gets in, it also can get out. But with most EIFS, when water gets in, it's trapped. That's why caulking around penetrations is so critical in an EIFS-sided building.

Mark Williams, an architect and building diagnostician, and Barbara Lamp Williams wrote the book on EIFS (ASTM manual 16, *Exterior Insulation and Finish Systems: Current Practices and Consideration*; out of print). "It is clear that faulty integration of the EIFS with other building-envelope systems—roofs, fenestration and typical residential-building features—is a prime cause of leakage problems," according to Mark Williams.

Although the industry claims that much of the water damage in North Carolina occurred because of cheap, shop-built windows that let water leak behind the insulation board, the Williamses found that even the best windows leak.

As part of his research, Williams mocked up an EIFS-sided house at his office in Maple Glen, Pennsylvania. On one window

**Water got through even this brand-name window. In an experiment, the sill was dammed with duct tape and water was poured in. After less than 1 minute, water began leaking through the sill-frame joint.**

(I won't name the brand, but it's a major window company, and the window wasn't a cheap one), Williams dammed up the sill using duct tape. Then he poured in a cup of water. After less than a minute, the water started dribbling through the joint where the sill is attached to the jambs (see photo above). Williams notes that builders can't be blamed for code-approved windows that leak water into an EIF system.

## EIFS Research Results Vary

Experts, most of whom have been hired to conduct research for one side or another in the EIFS controversy, disagree on what it will take to fix the problems. Williams, who is working for an insurance company that wrote policies for a lot of North Carolina

# An EIFS Retrofit for Leaky Windows

In a typical EIFS window installation, which has an exterior frame of built-up EIFS components, water that gets through the sill is held against the sheathing (see drawing at left). A retrofit design (drawing at right) includes metal pan flashing under the window. If water gets through the sill–jamb connection, it dribbles down the pan flashing, through a layer of metal or plastic lath, and out through another piece of flashing that exits in the shadowline of the lower window frame.

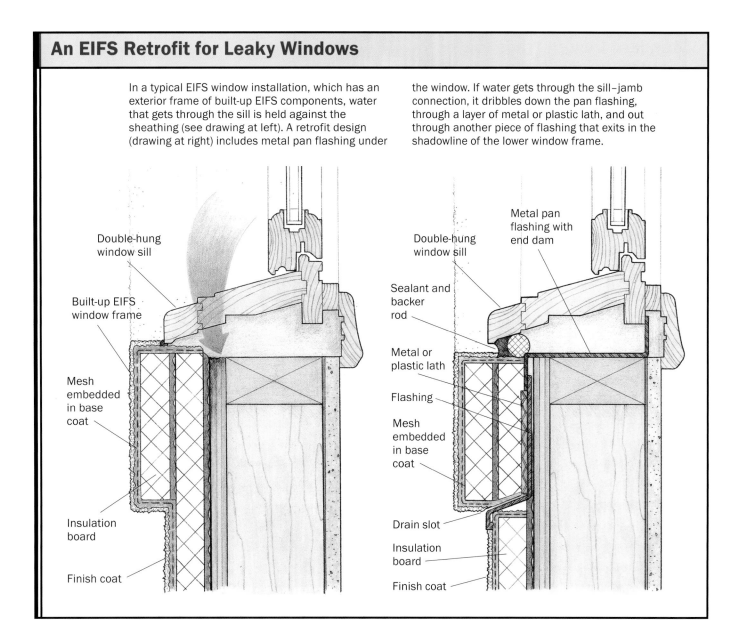

contractors, believes the industry must rethink its residential detailing. Dick Piper, a forensic engineer with R. J. Kenney Associates in Plainville, Massachusetts, believes the main problem "isn't with EIFS itself but with the way it's installed. Our position has been that standards for materials and applications need to be improved."

Building engineer Gary Zwayer of Wiss, Janney, Elstner Associates in Northbrook, Illinois, says, "My belief is that if you follow current EIFS manufacturer's details, you're going to avoid a lot of problems."

# Zeroing In on Windows

Assuming it's installed correctly, the actual EIFS surface is pretty good at staying put and at keeping out water. There have been problems with thin base coats or improperly installed insulation. But in general, a solid run of EIFS-covered wall has not been much of a problem.

Again, problems occur where EIFS is penetrated, especially at windows. So let's narrow our discussion to sealing around windows.

*For their part, builders need to understand that an EIF system isn't like ordinary siding.*

## Standard Design Traps Water That Leaks into Walls

Conventional EIFS detailing calls for insulation to be glued directly to sheathing. With no building paper or provision for drainage, water that leaks through the sill–jamb connection is trapped against the sheathing and causes rot.

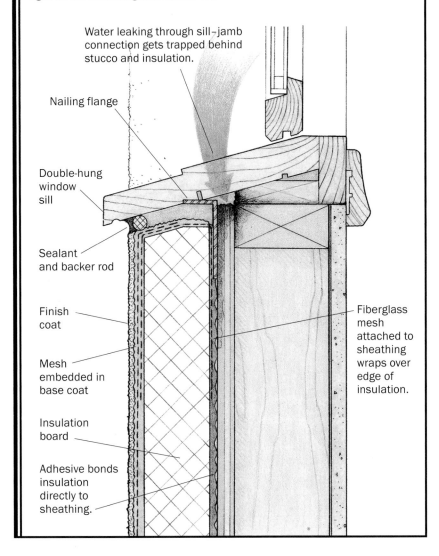

Water leaking through sill–jamb connection gets trapped behind stucco and insulation.

Nailing flange

Double-hung window sill

Sealant and backer rod

Finish coat

Mesh embedded in base coat

Insulation board

Adhesive bonds insulation directly to sheathing.

Fiberglass mesh attached to sheathing wraps over edge of insulation.

Manufacturers detail their window treatments a little differently, so I'll look at the way Sto details windows because Sto developed a water-drainage system in response to the problems in North Carolina. Sto is the second-largest EIFS company in the United States and the largest worldwide. Now based in Atlanta, it began in Germany.

In an ordinary Sto EIFS installation, the joint between the window and the siding is sealed with foam backer rod covered with sealant (see drawing at left). There is no window flashing shown in Sto's detail drawings. Now imagine this is the window Williams used in his mock-up. Any water that gets into the system through the sill–jamb connection drips behind the carefully laid backer rod and caulk seal.

Pan flashing installed under the window, along with some means of allowing water that leaks into the flashing to weep out, could have solved the problem, according to Williams, who designed a retrofit to channel water out of an EIFS-sided window assembly (see drawing on p. 41). Proper sealing and adequate flashing details at windows, doors, and roof–wall intersections would eliminate the need for a secondary weather barrier, such as building paper, according to Williams.

## How Things Are Changing

Sto Systems now has a modified version of its standard residential EIF system, which it has passed on to the building-inspection department in Wilmington. That system includes the use of a moisture barrier over the sheathing, flashing under windows, and what it calls a "drainage mat," or plastic lath, behind the insulation board (see photo and drawing on facing page). A second system Sto developed for North Carolina uses insulation board with grooves or channels cut in the back. Both are marketed in North Carolina.

**A watershed EIFS design.** Sto now offers this design in North Carolina: a layer of building paper, plastic lath for drainage, and the standard built-up insulation, mesh, and top coat. The wall terminates at the bottom into a plastic molding that contains weep holes.

Sto's Buck Buchanan, president of the EIMA, says Sto's standard system works just fine. "Outside of North Carolina, our approach will be this: If a builder is concerned about water, we've got these two systems, lath or grooved polystyrene. You still need proper details and windows that meet building code."

U.S. Gypsum (USG) pulled its brand of face-sealed, water-barrier EIFS off the market after problems in North Carolina. Now it's back as the Durock® Insulated System, which uses USG Durock cement panels as sheathing. However, the order of things in the USG system is a little different (see drawing on p. 44). Building paper goes over the framing, and on top of that is either expanded metal or plastic lath, then sheathing —if it's a humid climate—or just Durock sheathing. The rest of the installation is the same as in an ordinary EIF system. (Of course, cement panels cost more than either plywood or OSB. In southwestern Connecti-

## New Design Provides a Drainage Channel

With a drainage channel installed between the insulation and the sheathing (see drawing below), any water that gets into the wall hits the building paper and dribbles down and out through a weep screen at the base.

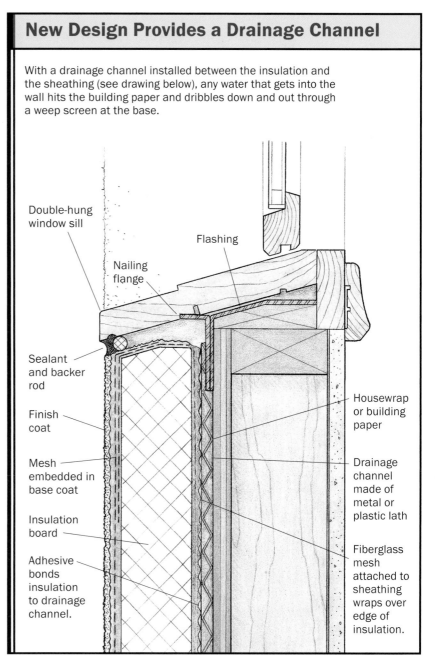

Double-hung window sill

Flashing

Nailing flange

Sealant and backer rod

Finish coat

Mesh embedded in base coat

Insulation board

Adhesive bonds insulation to drainage channel.

Housewrap or building paper

Drainage channel made of metal or plastic lath

Fiberglass mesh attached to sheathing wraps over edge of insulation.

> *Despite marketing that suggests the system is low maintenance, an EIFS-clad home needs regular inspection and care.*

cut, my local building supply sells ½-in. sheets of OSB for $7.15 per 4x8 sheet and ½-in. CDX plywood for $10.25 per 4x8 sheet. It charges $9.75 for a 32-in. by 5-ft. sheet of cement board.)

Parex is another manufacturer that has developed a drainage system for North Carolina. The Parex system is nearly identical to the Sto system shown in the photo on p. 43.

Although general EIFS installation guidelines are now available, the wisest stragegy for any builder interested in EIFS is to pay atten-

tion to the specific instructions provided by the EIFS manufacturer. Major manufacturers are represented by the EIFS Industry Members Association (see Sources). These manufacturers offer courses tailored to their systems.

Until the holes in the system are plugged, many experts suggest that a third-party inspection system be used to head off problems such as those that occurred in North Carolina. A report on North Carolina's problems issued in November 1995 by R. J. Kenney Associates suggests a range of precautions, including proper training, testing of the work, and "an inspector, either a trained clerk or a third-party inspector hired for the EIFS work."

## EIFS Isn't a No-Maintenance Siding

Despite marketing that suggests the system is low maintenance, an EIFS-clad home needs regular inspection and care. Sto recommends replacing damaged caulk every 3 years. It also suggests occasionally cleaning walls to remove pollution and not scarring or denting siding.

Pigment does go all the way through the textured finish coat, so it doesn't have to be repainted like many types of ordinary siding. However, it may have to be pressure-washed every few years, manufacturers suggest.

*Price estimates noted are from 1996.

---

*Steve Culpepper is former* Fine Homebuilding *magazine and book editor. He is currently editorial director for the shelter subsidiary of Sabot Publishing.*

## Cement Board Serves as Sheathing

The Durock Insulated System, produced by U.S. Gypsum in response to problems with its standard EIFS siding in North Carolina, uses cement board for sheathing and includes an interior water barrier installed directly over the studs.

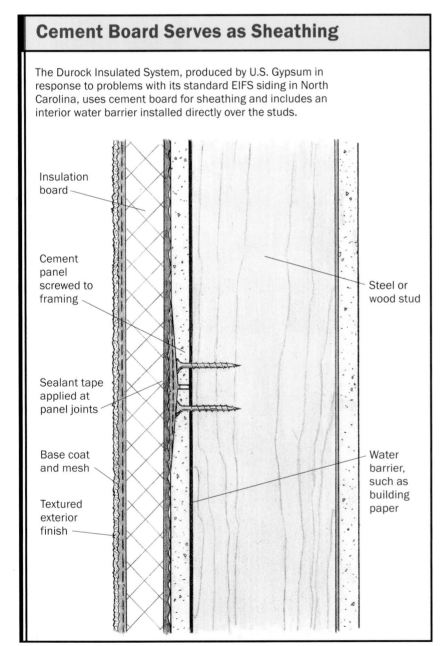

Insulation board

Cement panel screwed to framing

Sealant tape applied at panel joints

Base coat and mesh

Textured exterior finish

Steel or wood stud

Water barrier, such as building paper

# Flashing Brick-Veneer Walls

■ BY DOUGLAS R. STIEVE

**Flexible membranes** wrap all openings and bridge the gap between framing and metal flashing.

**Pricey and hard to work,** stainless-steel flashing resists galvanic corrosion around aluminum windows and doors.

**Durable, nonstaining** and easily worked, lead-coated copper is the author's recommended flashing in most cases.

**B**ricks and mortar may seem impervious to water infiltration, but they are not. Bricks and mortar themselves absorb water. Gaps between bricks and mortar can channel water into the wall of a house. Window and door openings invite in even more water. Most water infiltration can be prevented, but not all of it. So the real key to keeping house interiors dry is the way water flow is managed after it gets behind the brick.

Before the 1950s, most brick walls were thick, load-bearing walls consisting of at least two wythes of brick (a wythe is one thickness of brick). The bricks and mortar absorbed water, but usually not enough that it seeped inside. Since the 1950s, most residential masonry walls have been brick veneers, not load-bearing walls. Starting on the outside, modern walls usually consist of a single wythe of brick, a 2-in. drainage cavity, tar paper, wood sheathing and then a standard wood stud wall (see drawing on p. 46). Some people use housewrap in place of tar paper; but I don't because housewrap is not waterproof. Similarly,

> *The key to keeping house interiors dry is the way water flow is managed after it gets behind the brick.*

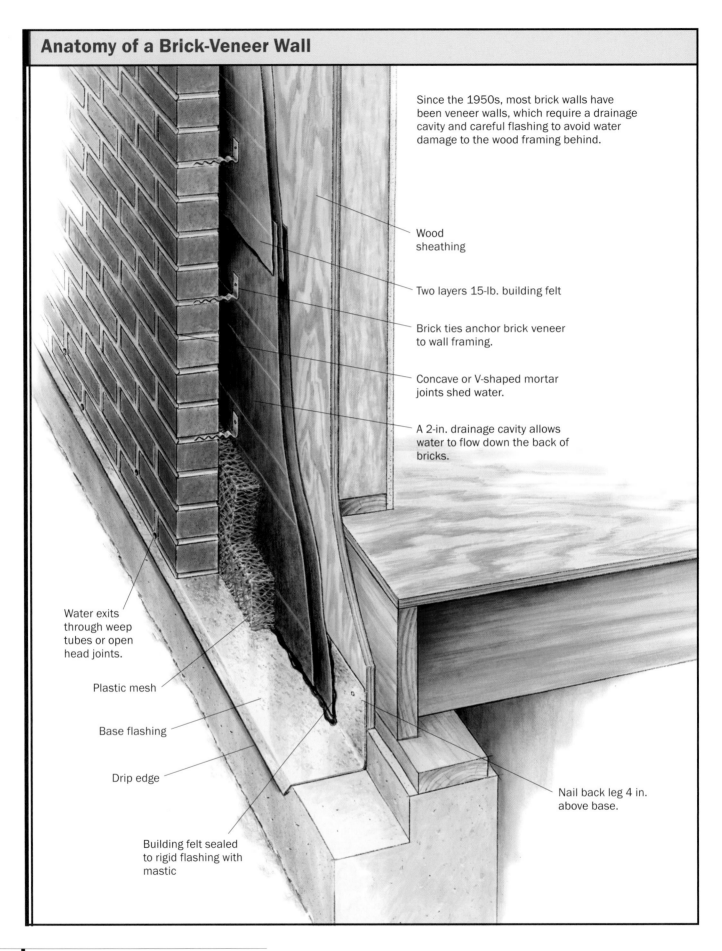

Since the 1950s, most brick walls have been veneer walls, which require a drainage cavity and careful flashing to avoid water damage to the wood framing behind.

Wood sheathing

Two layers 15-lb. building felt

Brick ties anchor brick veneer to wall framing.

Concave or V-shaped mortar joints shed water.

A 2-in. drainage cavity allows water to flow down the back of bricks.

Water exits through weep tubes or open head joints.

Plastic mesh

Base flashing

Drip edge

Building felt sealed to rigid flashing with mastic

Nail back leg 4 in. above base.

some people use only a 1-in. drainage cavity, but a 2-in. cavity helps prevent water from bridging across to the sheathing. The drainage cavity allows water to flow down the back of the brick onto flashings that direct the water back to the exterior.

At least that's the plan. Inadequate flashing details allow water to enter the framing, where it can rot the wood, soak the insulation, and stain the ceiling or walls. Brick-veneer walls directly above occupied sections of the home (where brick runs over a bay window, for example) are especially susceptible to problems. Any breakdown in the design or installation of the flashing and/or weep provisions of the exterior wall at these locations can cause disaster.

## Tar Paper and Flexible Membranes Protect the Sheathing and Openings

As an architect who specializes in fixing failing buildings, I've seen a slew of details that don't work. Here are some details I specify to ensure the longevity of these buildings.

I begin with two layers of 15-lb. tar paper on the sheathing, lapped to shed water downward (see drawing on facing page). The second layer provides redundancy in case of a tear that's not repaired before the masonry veneer goes up. Peace of mind and fewer

*Peace of mind and fewer callbacks are worth the cost of a second layer of tar paper.*

## Match the Mortar to the Brick

The right kind of mortar for the bricks is crucial. The two most popular forms of mortar are portland cement with lime and masonry cement. I recommend portland cement and lime because you know the constituents of the mortar and eliminate unnecessary additives.

Additives are often used to manipulate mortar's set time, porosity, or freeze–thaw durability. However, many of these additives are unnecessary, and using them improperly or in the wrong combination may damage the wall. To keep it simple, I usually recommend a type-S or -N portland cement and lime mortar. Because veneer walls don't carry structural loads, the mortar usually doesn't require great compressive strength. It does need the flexibility that types S and N provide. However, check local building codes before using the weaker type-N mortar.

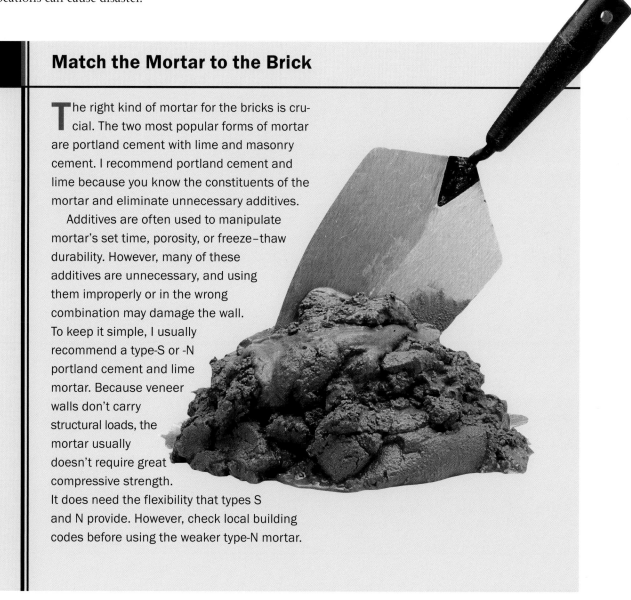

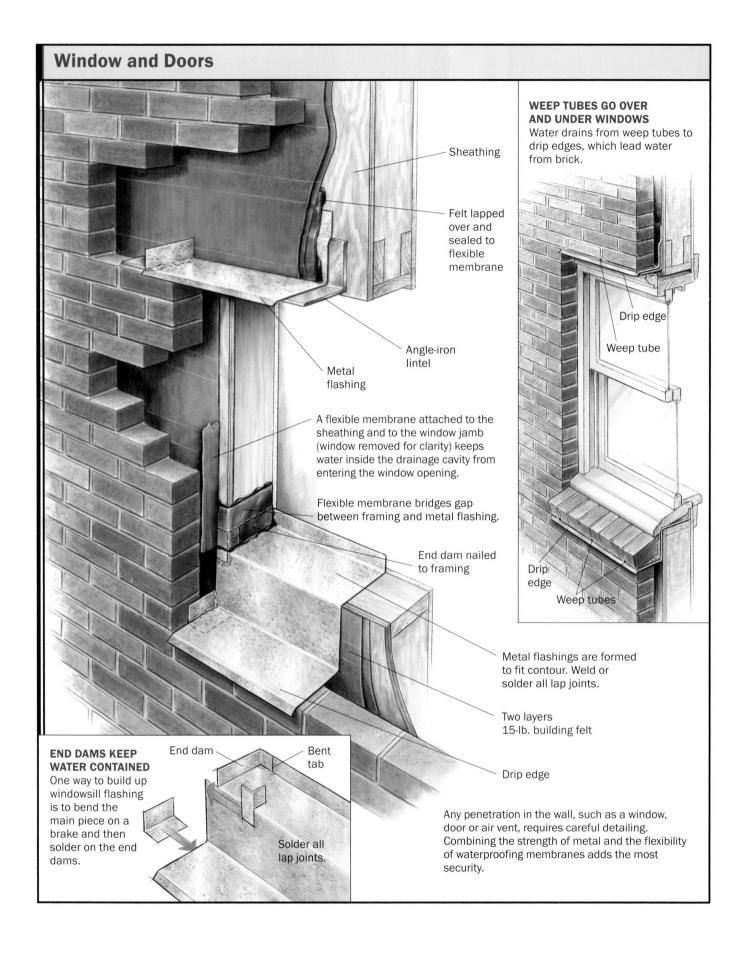

Sheathing

Felt lapped over and sealed to flexible membrane

Angle-iron lintel

Metal flashing

A flexible membrane attached to the sheathing and to the window jamb (window removed for clarity) keeps water inside the drainage cavity from entering the window opening.

Flexible membrane bridges gap between framing and metal flashing.

End dam nailed to framing

**WEEP TUBES GO OVER AND UNDER WINDOWS**
Water drains from weep tubes to drip edges, which lead water from brick.

Drip edge

Weep tube

Drip edge

Weep tubes

Metal flashings are formed to fit contour. Weld or solder all lap joints.

Two layers 15-lb. building felt

Drip edge

**END DAMS KEEP WATER CONTAINED**
One way to build up windowsill flashing is to bend the main piece on a brake and then solder on the end dams.

End dam

Bent tab

Solder all lap joints.

Any penetration in the wall, such as a window, door or air vent, requires careful detailing. Combining the strength of metal and the flexibility of waterproofing membranes adds the most security.

callbacks are worth the cost of a second layer of tar paper.

Tar paper alone, however, isn't enough around windows, doors, and mechanical penetrations. Metal flashing and flexible waterproofing membranes are needed to seal these vulnerable openings. These flexible membranes are composite materials, usually containing rubberized asphalt with a cross-laminated polyethylene-film reinforcement. One such product is Bituthene® 3000 by W. R. Grace. Some people use Grace's roofing underlayment, Ice & Water Shield®, for flashing, but Bituthene 3000 is a little thicker and stronger. Some manufacturers offer flexible through-wall flashing to take the place of rigid metal. I don't recommend this because metal is more durable and more easily formed into a drip edge.

# Rigid Flashings Channel Water Out

Combining rigid-metal flashings with flexible waterproofing membrane is the best way to divert water from wall penetrations (see drawing on facing page). Metal stands up well to the installation of the window or door and provides a clear exit drain.

Metal flashings come in a variety of forms. I prefer 16-oz. lead-coated copper because it won't corrode. Lead-coated copper is easily formed with a metal brake either in a shop or in the field, and watertight seams are easily soldered in the field. The lead coating prevents the copper base material from staining the masonry. Other rigid flashings include kraft paper–laminated copper, type 304 stainless steel, aluminum, and polyvinyl chloride (PVC). The latter two require a caution. Prepainted aluminum is common, but unpainted aluminum will corrode if set in wet mortar. PVC tends to turn brittle and break down over time.

Bending up the backs and sides of the metal flashings to form vertical back legs and end dams prevents water from spilling off the ends of the flashing and down under the windowsill or into the space between the window frame and the adjacent studs. All seams should be soldered or welded watertight. The angle-iron lintel that's typical above a door or window (see drawing on facing page) should not be relied on to serve as the rigid flashing. The lintels have no end dams to keep water from flowing over the sides, and they tend to corrode.

Because metal flashing can be difficult to install tight to adjacent wood studs and sheathing, flexible waterproofing membrane should be adhered to the studs and sheathing, underneath the tar paper, and lapped into the metal pan. The window or door should be shimmed slightly off the bottom flashing to allow water to exit between the bottom of the window or door and the flashing.

# Water That Gets behind the Brick Must Exit at the Base of the Wall

At the bottom of each wall section, the tar paper should lap onto a metal flashing, just as at the windows, adhering to the flashing with the same waterproofing mastic used for flexible membranes and touching the horizontal leg of the metal. The vertical back leg should be nailed directly to the sheathing at least 4 in. above the horizontal leg so that the nail penetration is well above the level of any collecting rainwater. The horizontal leg extends completely to the exterior of the wall and forms a drip edge to prevent water from traveling around the edge of the flashing and back into the wall cavity.

## TIP

*Shim the window or door slightly off the bottom flashing to allow water to exit between the bottom of the window or door and the flashing.*

## Sources

**Advanced Building Products**
(800) 252-2306
www.advancedbldg
products.com

**American Society for Testing and Materials**
(610) 832-9585
www.astm.org

**Brick Industry Association**
(703) 620-0010
www.bia.org

**Hohmann & Barnard Inc.**
(631) 234-0600
www.h-b.com

**W.R. Grace**
800-444-6459
www.grace
construction.com

Metal flashing should wrap continuously around all inside and outside corners (see drawings below). Corner flashings can be prefabricated in a shop, but because no building corners are ever perfect, I prefer making the corner flashings in the field so that adjustments can be made for construction tolerances.

To allow water to drain, weep holes should be made in the brick veneer at the level of the rigid flashing, both at the base of the wall and above windows and doors (see top right inset drawing on p. 48). Open head joints (the vertical joints between two bricks in the same course) provide an effective weep system, but they also create a wide-open path for wind-driven rain to blow into the wall. Oiled rods that are laid in with the mortar and then removed after the mortar cures form weep holes that don't present such a big target for rain.

Plastic weep tubes spaced at 16 in. o.c. (or every other head joint) also work well. Installing two rows of weep tubes, with the

## Base-Flashing Corner Details

Corner flashings can be formed in the field from lead-coated copper. The flashings should be nailed to the sheathing at least 4 in. up the back leg, and building felt or flexible waterproofing membrane should overlap and be adhered to the metal base flashing.

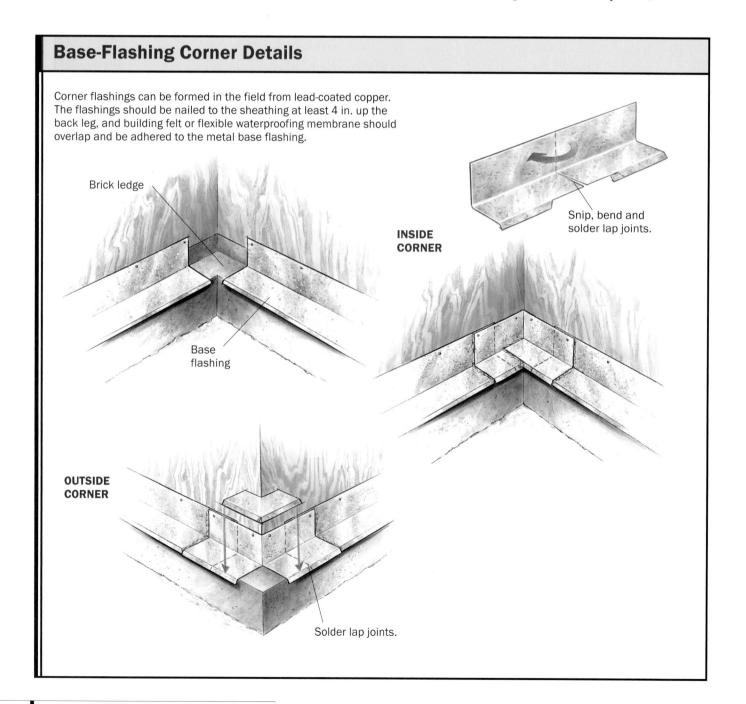

Brick ledge

Base flashing

INSIDE CORNER

Snip, bend and solder lap joints.

OUTSIDE CORNER

Solder lap joints.

## Some Types of Brick Call for Wetter or Drier Mortar

All bricks absorb water through microscopic pores and do so at different rates. Knowing the rate at which bricks absorb water from mortar, their initial rate of absorption (IRA), can reduce the amount of cracking and separation that occurs later.

If the brick absorbs water too quickly from the fresh mortar, little water will be left to properly hydrate the portland cement in the mortar, and subsequent cracking within the mortar may occur. If the brick absorbs too little water, however, a weak bond forms, causing the mortar to separate from the brick.

Mortar recipes can be adjusted, if necessary, to accommodate the brick's IRA, which can be obtained directly from the brick manufacturer. A brick with an IRA above 25 g per minute per 30 sq. in. is considered very absorptive, and an IRA below 5 g per minute per 30 sq. in. is considered low. For a very absorptive brick, make one of the following adjustments:

- Add a little more water to the mortar and wet the bricks before installation (except in cold weather).
- Add more hydrated lime within the limits of ASTM C270.
- Use a workability-enhancing admixture that conforms to ASTM C1384.

For a low IRA, the brick manufacturer's recommendations may include one or both of the following adjustments:

- Reduce the amount of water in the mortar.
- Keep the brick dry before installation.

For more details on bricks, mortars, and techniques, contact the Brick Industry Association and the American Society for Testing and Materials.

A dense brick with a less porous surface absorbs less water.

A less dense, more porous brick can take a wetter mortar.

## Catch Mortar Droppings to Keep Weep Holes Clean

Several manufacturers sell plastic mesh to break up and deflect mortar droppings before they reach the bottom of the cavity and block weep holes. The meshes come in different heights and thicknesses, and are dropped into the cavity as the first bricks go up.

**Mortar Break** by Advanced Building Products

**Mortar Net** by Hohmann & Barnard Inc.

second row a single course above the first, provides some redundancy in case the lower section of the wall fills with mortar droppings. Plastic weep tubes should be laid in the head joints before the brick is pressed into place. Shoving weep tubes into an already formed joint fills them with the mortar.

# Properly Laid Brick Sheds the Bulk of the Water

Because the mortar in the head joints isn't compressed as much as that in the horizontal bed joints, head joints are not always completely filled with mortar and thus can invite water infiltration. To be as watertight as possible, joints between bricks should be completely filled with mortar. Mortar should be packed into the head joints without spilling into the drainage cavity.

Mortar droppings at the bottom of the drainage cavity often block weep holes. There are a few techniques to limit mortar droppings. Mesh materials inserted into the drainage cavity catch the droppings before they get to the bottom. If enough mortar accumulates, however, it covers these products and blocks the drainage cavity. Back-beveling the mortar for bed joints (striking it so that it angles away from the cavity before stacking on the next course) makes for a smaller amount of mortar droppings in the first place.

*Douglas R. Stieve* is an architect with Wiss, Janney, Elstner Associates in New York City who specializes in the diagnosis and repair of building envelop failures.

# Installing Manufactured-Stone Veneer

■ BY VLADIMIR POPOVAC

**Y**ears ago, I was driving down the road and happened to glance over at a construction site. Two masons perched on scaffolding 25 ft. in the air were setting stone on the outside of a chimney. They had completed about one jagged foot of stone all around the top of the chimney, working down instead of up.

I couldn't believe my eyes. Ordinarily, I'd stop to find out what was going on. But I was too stunned to react, and I just kept driving. Later, I discovered their secret: They were installing manufactured stone.

## Manufactured Stone Looks Like the Real Thing

Most people have trouble discerning manufactured stone from natural stone, even up close. Even experienced masons can have difficulty distinguishing the two. Phony stone, as some masons call it, looks like the genuine article in both shape and color.

So what's not to like about the stuff? First, lightweight manufactured stone lacks the heft of natural stone. And to the experienced hand and eye, it doesn't have the same solid feel or depth of color as rock that has formed over millions of years. It's also applied more like tilework than stonework.

Natural stone has color throughout each piece, but manufactured stone's beauty is skin deep. The color is just on the outer surface, which makes cleanup a bit more problematic, something I'll get to later.

But manufactured stone does have advantages, the biggest of which is that it can be installed from the top down. Most manufacturers' instructions state that it can be installed in either direction, but working in the reverse of standard masonry logic makes cleanup easier because you are not dropping mortar all over the completed work.

Compared with natural stone, manufactured stone cuts more easily and more predictably. Made from cast concrete, manufactured stone has an even consistency in each piece, making it easy to shape and fit.

*Manufactured stone's biggest advantage is cost.*

**Take a closer look. From here, it's tough to tell that the stone on the foundation of this house is actually lightweight concrete cast and colored to look like the real thing.**

Manufactured stone's biggest advantage is cost. The clients on this project wanted natural river stone for both an inside hearth and for the veneer on the foundation and chimney outside. When the price came in more than they wanted to spend, they decided to use manufactured stone outside, which cut the cost in half. The money they saved let them use the real McCoy inside where the stone would be seen much more closely. Manufactured stone comes in a variety of shapes and colors, so it was no problem duplicating the river-washed look of the Canadian stone they had chosen.

# Before the Stone Goes On

Manufactured stone has to be applied over a cementitious surface. If it is going over existing masonry, such as concrete or block, the surface has to be clean and solid. No wall ties, anchors, or lath is needed. And because of its light weight, no support—such as angle irons or a veneer shelf cast in the foundation—is required.

If the subsurface isn't cementitious, such as drywall, plywood, or rigid insulation board, you must make it so. The goal is not to build structural support but to create a solid surface that won't move or flex.

Most of the crawlspace walls of this house were wood-framed with plywood sheathing. In situations such as this one, my partner, Marshall Dunn, and I begin by tacking up a moisture barrier. Kraft paper or 15-lb. felt paper can be used. This barrier keeps the subsurface from sucking moisture out of the mortar, causing it to cure too rapidly.

Next, we nail expanded metal lath flat over the entire surface with galvanized roofing nails driven into the framing. These nails are the only mechanical anchors holding the manufactured stone to the wall, so don't skimp on them.

Metal lath can be cut with shears, but cut edges are sharp and can slice your hands if you don't wear gloves. Instead, we use a small circular saw with a diamond blade to make fast cuts while keeping fingers away from the sharp edges (see top photo on facing page).

# A Mud Layer Finishes the Substrate

Next, we spread a thin layer of mortar over the paper and lath. We use a standard mix: 3 parts plaster sand (finer than concrete

A saw keeps fingers away from sharp edges. Metal lath for the substrate can be cut with shears, but cutting with a diamond-blade saw is quicker and safer.

sand), 1 part portland cement and 1 part clay. Lime may be substituted for clay, but either provides the bonding power that makes the mix stick to the wall.

We usually spread the mortar with a brick trowel (see bottom left photo). The trick is to get the mortar on the wall quickly and to keep it flat. Often, we heap a pile of mud on a piece of plywood next to the wall and push the mud directly onto the wall. While the mortar is still slightly wet, we furrow the surface with a toothed trowel to maximize the mortar's grip (see bottom right photo).

We let the prepared surface set at least overnight, and after that point, the prepped wall is treated the same as any other cementitious surface. The next day, before we start installing the stone, we soak the wall with either a sponge or a hose. A dry wall, especially on a hot day, can rapidly suck the moisture out of the wet mortar, weakening the bond between the stone and the wall.

*The trick is to get the mortar on the wall quickly and to keep it flat.*

Smearing the mud. Piled on a sheet of plywood, mortar is pushed up onto the wall, where it is spread in a thin layer over the metal lath.

A textured finish provides better grip. A notched trowel creates a furrowed surface that is better for bonding the manufactured stone to the wall.

**Pick a stone, any stone.** Manufactured stone is laid out on plywood to keep it clean while the masons choose pieces to fit spots on the wall.

# Two Shapes: Flats and Corners

We begin installation by laying out a large number of stones close to where we'll be working. If we're outside, we set them on tarps or sheets of plywood (see photo at left). Having many stones laid out beforehand makes finding the right stone for a particular spot much easier.

Manufactured stone has to be kept clean and dry before it is installed. Mortar doesn't bond well to dirty or wet stone, and stains in manufactured stone are usually permanent. So at the end of each day, any unused stone is returned to its crate and covered.

Manufactured stone comes in two basic shapes: flats and corners. The flats have a front surface that resembles a stone in outline, convolutions and color. But they are only 1 in. to 3 in. thick, and their backs are flat. Corners have their backsides molded into a right angle. Because corners are harder to shape and give you fewer options, we usually begin with the corners, letting them define the border of the field.

# Cutting and Fitting the Stone

The first piece we set is a top cornerstone (see top right photo). We use the same mortar mix to set the stones as we used for the subsurface. Once the top piece is scribed, cut, buttered, and tapped into place, we move to the next piece down.

The two arms or wings on each cornerstone vary in length, in width and in thickness. So as I move down the corner, I alternate the orientation of the wings whenever I can, to produce a more natural look (see bottom right photo). I also alternate color whenever possible.

With corners or flats, we start with a stone that looks like a close fit with its neighbors, then shape it as necessary. We

**Start in a corner.** Corners are harder to shape and fit, so the first stone to go on is a top corner piece (photo above). Color and shape are alternated for a more natural look as the corner pieces are installed (photo below).

determine where to chip by holding the new rock in place. We sometimes mark the cut with a pencil (see top right photo), but usually the eye remembers where and how much to cut. And if the first shaping is not perfect, you just keep shaping until you're satisfied.

I like to use a masonry hammer to do the chipping, but my partner shapes the stone with the edge of his brick trowel. Either way, support the stone while you're striking it; manufactured stone is fragile before it's wedded to the wall. Cradle the stone in one hand against your body, and attack the area to be removed from back to front as if you were whittling. Be patient; remove the unwanted material a little at a time. (see bottom right photo).

Although we prefer to shape the stone by hand, occasionally we need to remove a larger section of stone than is feasible by chipping. In this case, we use a 4-in. hand-held grinder equipped with a diamond cutting wheel. It's fast and accurate. You can even soften and slant the cut, which helps to hide it. The only drawback is the dust produced. We always wear masks when using the grinder, but we prefer to shape the stone by hand.

## Cutting Manufactured Stone

**B**ecause manufactured stone has color only on the surface, we try to minimize the visibility of the cuts. Manufacturers suggest that if the stone to be set is above eye level, then you should shape the stone from the top and vice versa. But taking special care while cutting and then using creative mortar techniques can hide most cuts.

**Isn't this guy working backward?** Installing manufactured stone from the top down keeps excess mortar off completed work. Here, a piece is scribed for a fit.

**Shaping manufactured stone is a breeze.** Cast-concrete manufactured stone has an even consistency, making it easy to shape with either the edge of a trowel or a mason's hammer. An electric grinder can also be used.

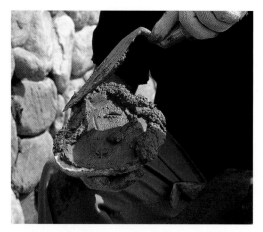

**Buttering the backs.** A thin layer of mortar is spread across the back surface of the stone, and a ridge around the edges helps fill joints between the stones.

**A light tap sets the stone.** After a stone has been pressed into place, a light blow with a soft mallet ensures that the stone will stay put while work continues.

## Buttering the Backs

When we're satisfied with the fit of a stone, we spread a thin bonding coat of mortar over the back of the stone, pressing the mortar firmly into it to ensure good adhesion. Next, we run a small ridge of mud ½ in. high around the edges of the stone (see top left photo). The extra mud oozes out to fill the joint when the stone is pushed into place. If we're installing stone with a tighter pattern, less mud is needed around the edges.

The back of the stone must be bonded to the wall, but too much mortar can spell trouble. First, the excess could run onto the face of the stone, adding to your cleaning chores. But even more serious, too much mud adds extra weight, and the stone will fall as soon as you turn your back or tap the next stone into place. Practice and experience dictate the right amount of mud.

We hold the buttered stone about 1 in. away and 1 in. below its final position, then slide it into the wall and up, wiggling the stone as it slides to spread the mortar uniformly across the back of the stone and into the joints above. While holding the stone in place, we tap it a couple of times with a soft mallet and let go (see top right photo). The stone should now be set.

The large, uneven joints of this veneer made it necessary to add mortar occasionally to fill in the joints as the installation proceeded. We use partially dry mortar that still adheres well but won't smear and stain the stone. Grouting mortar can be either put in by hand or squeezed out of a grouting bag.

## Manufactured Stone Needs Special Care for Cleanup

Before we get too many stones installed, we stop to clean up, especially if it's a warm, dry day when the mud sets up more quickly. Cleaning manufactured stone takes much more care than cleaning natural stone. With the latter, you can scrub as hard as you'd like, using abrasives, wire brushes, and acid without altering the hard surface and body of stone.

But phony stone won't stand up to the same kind of abuse. Any mortar stains must be cleaned within the first few hours, or they become permanent. Acid and abrasives that can erase the colors from the surface are absolutely verboten. Again, working from the top down and using the right mix should help keep messiness and cleanup to a minimum.

# Concrete Cast in Stone

**Cobblefield**

**European castle stone**

A few of manufactured stone's many faces. Because it is a cast-concrete product, manufactured stone can resemble many different types of stone, each of them in a variety of colors.

**Ledge stone**

**M**anufactured stone is a lightweight, nonstructural alternative to natural stone made by several different companies. Although the process may vary slightly from company to company, most manufactured stone is made by pouring lightweight concrete mix into molds made from actual stones.

Pigment placed in the mold before the concrete is poured gives each stone its color, and most companies offer many types of stone in a variety of colors. The stone featured in this article is River Rock, Lakeshore Blend from Cultured Stone®.

Ordering manufactured stone is easy. First, figure the square footage minus openings such as doors or windows. Next, add up the linear feet of the outside corners. Each company supplies figures for how many square feet each linear foot of cornerstone covers for a variety of stone. The square footage of the corners is subtracted from the total for the final figure.

Because of its forgiving nature, there is little waste with manufactured stone, but it doesn't hurt to order a little extra.

**Coral stone**

**Stream stone**

**Cleanup begins with a small stick.** When mortar has just begun to set up, a small soft-wood stick is used to remove excess. The pointed end rakes the joints between stones to the right shape and depth.

**A gentle hand with a soft brush.** The color on the surface of manufactured stone can be damaged with abrasives or acid, so excess mortar must be brushed off lightly with a soft brush.

The ideal time to clean the stone and rake the joints is as soon as the mortar dries but before it hardens. Exactly when this ideal moment arrives depends on a number of factors, including the temperature, the humidity, the location of the sun (or the shade), the wetness of the wall, and the stiffness of the mud. But usually it occurs 1 hour to 3 hours after the stone is set.

Even if you've been a little sloppy in spots, resist the temptation to clean the stone until the mortar has set up. Cleaning the stones when the mortar is still wet can smear the mortar and stain the stone.

The first step in cleanup is raking the joints. When the mortar has dried sufficiently, we find a piece of soft wood, such as a redwood or cedar shim, about 6 in. long and the width of a finger (see top photo). With this stick, we flick off globs of mud that are perched on the stones and rake out the joints to the desired depth.

As this simple stick is used, it shapes itself into the perfect tool. We guard the sticks we use and hide them in places we're sure to remember for tomorrow or next week. But, of course, we never find them again.

The deeper you rake the joints, the smaller they get and the tighter they look. Just be sure not to expose the stone's cut edges.

Once the surface mud is flicked off and the joints are properly raked, we brush the stones with a soft, dry or barely damp brush (see bottom photo). Brushing takes away all the tiny, lingering particles of mortar. After brushing, the stonework should be almost totally clean and the joints should be smooth and uniform in depth.

The final step is washing, which is actually too strong a term. We wipe the stone lightly with a sponge just damp enough so that water isn't squeezing out (see photo on facing page). Any remaining stains should lift right off. If a stain requires heavy rubbing, you've waited too long, but go ahead and rub. Tomorrow will be too late.

*Vladimir Popovac* and his partner, Marshall Dunn, are masons in Sonora, CA.

**Sources**

**Century Stone**
(780) 447-2122
www.cci-industries.com

**Classic Stone**
(800) 808-0027
www.classicstone.com

**Coronado Stone**
(800) 847-8663
www.coronado.com

**Crown Hill Stone**
(800) 295-1120
www.crownhill.com

**Cultured Stone**
(800) 255-1727
www.culturedstone.com

**El Dorado Stone**
(800) 925-1491
www.eldoradostone.com

**Hammer Stone**
(800) 246-1543
www.hammerstone.net

**J&N Stone Inc.**
(800) 321-1372
www.jnstone.com

**Tejas**
**Textured Stone**
(866) 578-5616
www.tejasstone.com

**A damp sponge completes cleanup.** Any light stains or mortar left after brushing is wiped from the stone with a sponge that is barely damp.

# Laying up Stone Veneer

■ BY STEVEN SNYDER

On a trip through Bucks County, Pennsylvania, in the early 1980s, Brenda and Buddy Williams noticed a stone house that soon became something of an obsession. During subsequent visits over the next several years, they studied the house, which is called Burgess Lea. Built in 1689, the house is an exquisite example of early Quaker architecture, and it had been brought back to museumlike quality by its owner.

I had become the resident stonemason at Burgess Lea, working there every year on jobs that took from a few weeks to a month or two. In 1989, the Williamses approached me with their plans to build a house in Mississippi patterned after Burgess Lea. They asked for my help in understanding the stone details of the house and how they could reproduce them economically. Fortunately, I had plenty of experience in reproducing period stonework in 6-in. and 8-in. veneers. Because of the huge economy of labor and materials in veneer construction, not to mention the insulating advantages of frame construction, the Williamses decided that veneer construction was the most reasonable way to go. An 8-in. veneer on a 2x6 frame could approach the 18-in.-thick stone walls of Burgess Lea.

After showing me their blueprints and thanking me for my input, they prepared to leave. On her way out, Brenda turned to me and asked if I would consider doing the stonework. I thought she was joking, but two years later, I was on my way to Mississippi.

( Note: Since this article originally appeared, I have stopped using portland cement–based mortars, and now feel they are inappropriate in most applications. I currently use hydraulic lime-based mortars on the preservation, restoration, and reproduction work my stonework company focuses on.)

## Salvaged Stone Doesn't Go as Far

The sandstone of the Williamses' house was to match that of Burgess Lea's. Because freshly quarried stone would not have matched the stones used on Burgess Lea, we used salvaged stone from Pennsylvania. As I collected the stone I'd need for the job, I considered its color, size, and shape. An old barn, a dismantled prison, stone quarries, and fields all yielded parts of the new house.

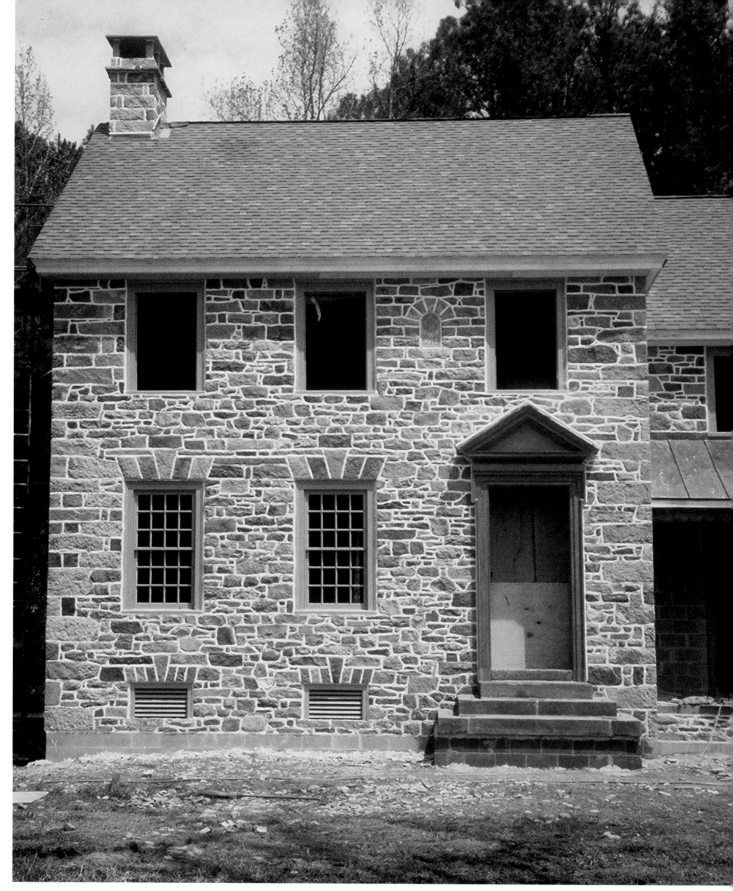

**Salvaged stone makes a new house look old.** Patterned after a 17th-century house in Bucks County, Pennsylvania, this Mississippi home sports 45 tons of stone facade. Because freshly quarried stone would have looked too new, Pennsylvania sandstone was salvaged from an old barn, a dismantled prison, stone quarries, and fields.

A ton of stone will cover 25 sq. ft. to 30 sq. ft. in an 8-in. veneer. Most quarries that sell building stone cut or split the pieces into 4-in., 6-in., or 8-in. widths for veneers. These stones don't produce much waste, so 30 sq. ft. per ton should be adequate for an 8-in. veneer. But because salvaged stone must be trimmed to width, it produces more waste, and I counted on covering only 25 sq. ft. per ton. My job was to veneer the facade of the main house, the center addition, and the foundation around the kitchen wing. The rest of the siding would be wood. For a little more than 1,100 sq. ft. of stonework, I needed approximately 45 tons of stone.

Some of the stone details could be reproduced using templates. Working with my longtime friend and helper, Tom Ashburn, I copied the arch stones and the shape of the date stone from Burgess Lea, then produced most of these pieces at my yard in Point Pleasant, Pennsylvania. Then the stone was wrapped in stretch film on pallets and loaded onto two tractor trailers.

At the building site, the first job was to make sure we'd have a dry place to work. Mississippi had been getting heavy rain for several weeks by the time we arrived in January, so I tented the work area. Using 2x4 rafters attached to a plate at the eaves, I enclosed the front of the main section of the house, running the 2x4s about 15 ft. out from the house (see photo below). The ends of the rafters rested on a temporary stud wall built near the house. Three pulleys along the top of this open roof made retracting a tarp simple. Even during heavy rains, work proceeded comfortably. The tent was more than worth the effort.

## Accurate Reference Lines Mean a Straight Wall

With a dry work area ensured, the next step was to establish accurate reference lines to use as we laid up the stone veneer on the main part of the house. It's easier and more accurate to establish the plane of the wall independently rather than take measurements from the sheathing. Even a slightly racked or out-of-plumb frame can cause some awkward problems for the stonemason later on.

I began setting these reference lines at the soffit. The innermost trim board should be hung so that the distance between its inside edge and the sheathed wall equals the thickness of the veneer, in this case 8 in. After establishing this point at each corner of the house, we pulled a line between the two points so that the trim would create a true line across the front of the house. Carpenters then hung the soffit and attached the

**Make a tent to stay dry.** A simple 2x4 frame covered with a tarp allows work to move ahead, even during rainy weather. Line run through pulleys at the eave of the house make raising and lowering the tarp simple.

**Setting reference lines.** The author's system for ensuring flat, plumb walls starts with reference lines hung at the top two corners of the wall. The trim piece just below the soffit is installed so that its inside edge is 8 in. from the sheathing. A plumb line is dropped to the foundation from this point to mark the corner.

**The horizontal string moves up and down.** The author set two vertical stringlines at each corner. Between the inside lines he stretched a horizontal line that could be moved up and down on the vertical lines and that established the stone wall's plane. Galvanized metal ties then are nailed into studs.

trim along its lower edge. On the back edge of this trim piece at both upper corners (8 in. from the face of the sheathing), we set finish nails. From those points we dropped plumb lines to the foundation (see top left photo) and set masonry nails to receive the lines. (Experience suggests this not be done on a windy day!)

We pulled two separate lengths of braided nylon mason's line taut between top and bottom reference points at each corner. These two sets of lines marked the vertical face of the wall at each corner of the house.

Next, we ran two horizontal lines across the face of the house. These horizontal lines were pulled fairly tight and were tied to one of the vertical lines at each corner (see top right photo). The horizontal lines could slide up and down on the vertical lines, providing movable reference lines. The plane defined by these lines as they are moved up and down is truly flat and plumb. The remaining vertical line at each corner, to which horizontal lines had not been attached, established a straight, plumb reference mark.

# Stone-Veneer Walls Need Beefy Foundations

The foundation was complete by the time we arrived, but it's important to note that stone-veneer walls require a lot of support. In normal frame construction, the footing is several inches wider than the foundation wall that will carry the weight of the house. For a house that is to be veneered in stone, the foundation needs to be wide enough to carry the frame and the stone veneer.

Ideally, a 16-in.-wide foundation wall on a 2-ft.-wide footing is built up to within 6 in. of final grade. At that point, the wall would be stepped back 8 in. to create a shelf to carry the weight of the wall.

Stone is heavy, so plan the foundation accordingly. An extra-wide footing supports an extra-wide foundation wall that will carry both the framed wall and the stone veneer. The 8-in.-wide shelf is capped with solid concrete block 4 in. thick, providing a strong surface for the start of the stone.

# Setting Windows and Installing Wall Ties

Windows were to be recessed 1½ in. from the face of the finished wall. Carpenters used 2xs to box out the rough openings in the frame wall and then hung the finished windows in the openings at the correct setback from the reference lines (see photo at right). Because the windows were recessed, the horizontal reference lines could pass freely in front of them. Setting windows this way ensures that all windows will have the same setback from the finished stonework. Plywood protected the windows during construction.

With the windows set, we were ready to prepare the surface against which the stonework would be laid. The 2x6 frame wall had been sheathed in ½-in exterior plywood, and we covered that with 15-lb. roofing felt to provide a moisture barrier. We started at the bottom and moved up, overlapping each course of felt by 4 in. to 6 in.

Next came the anchoring system for the stone. We located the studs at the bottom of the wall and, using a 4-ft. level, drew plumb lines on the roofing felt up to the soffit, marking each stud. Heavy-gauge galvanized wall ties were then anchored to the frame using 2-in. galvanized roofing nails. The wall ties should be applied 16 in. o.c. both vertically and horizontally. Most of the wall ties are hung before laying the stone, and the ties generally fit between courses of stone. But because corner stones tend to be bigger and less regular, I set wall ties there as I went along to make sure the wall ties fell in corner joints.

# Disguising Veneer Corner Returns

One of the main objectives on this project was to create the look of solid-stone construction. But when a wall has been veneered, it is common to see a consistent 4-in., 6-in., or 8-in. return at the corners

**Getting windows away from the wall.** Windows would be recessed too deeply if they were installed directly in the wood-frame wall. Instead, carpenters built extra boxes out of 2x material, nailed the boxes into the rough-frame openings and then set the windows outside the boxes. Reference lines ensured a uniform setback for windows in the finished stone-veneer wall.

running straight up the building. Corners in a solid-stone house would be less uniform. Disguising this return would help us achieve the look we wanted.

At the southwest corner of the house, the stone wall would join wood siding. I wanted the corner board and the siding to come out nearly to the outside corner of the stone, thus covering most of the 8-in. thickness of the veneer. To accomplish this, we extended the sidewall at the corner (see drawing on p. 68). After beveling one edge 60°, we lag bolted a 2x8 to a double 2x4 nailer set on

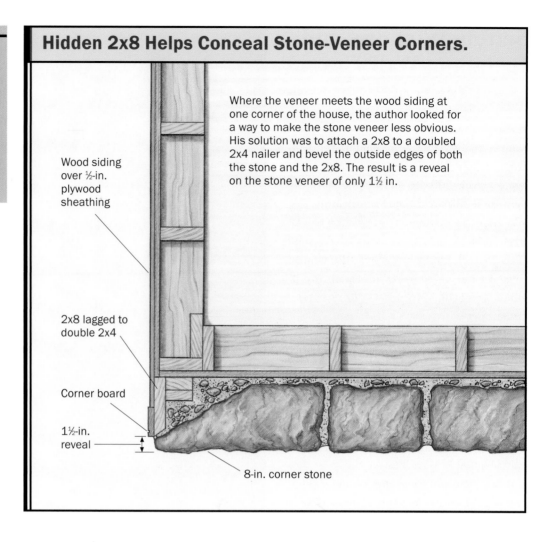

## Hidden 2x8 Helps Conceal Stone-Veneer Corners.

Where the veneer meets the wood siding at one corner of the house, the author looked for a way to make the stone veneer less obvious. His solution was to attach a 2x8 to a doubled 2x4 nailer and bevel the outside edges of both the stone and the 2x8. The result is a reveal on the stone veneer of only 1½ in.

Wood siding over ½-in. plywood sheathing

2x8 lagged to double 2x4

Corner board

1½-in. reveal

8-in. corner stone

the front wall, extending the sidewall to within 1½ in. of the face of the stone veneer.

The outside edges of the corner stones were chiseled back 30° and laid to the 2x8. The result is a 1½-in. reveal at the corner. The joint between stone and wood can be caulked.

## Mortar Should Have Lots of Body

With our lines laid out and the surface prepared, we were ready to begin laying stone. Whether building a solid-stone wall or applying a stone veneer, I prefer a sticky mortar with a lot of body. I see mortar not as a bonding agent but as a stable fill that accommodates the irregularities of the stone. The old adage is "Mortar doesn't hold stones together; it holds them apart." Good stone-laying techniques, not strong cement mortar, will result in a solid, long-lasting stone wall.

Lime adds body to mortar. My mix includes 1 part portland cement, 2 parts hydrated lime and 6 to 9 parts sand. For the mortar on this project, we used a sand with a small-pebble aggregate commonly known as concrete sand. The coarse sand combined with the high lime content to provide the body needed to support the stone as it was placed.

# Starting Walls with Basement Vents

The louvered basement vents on the main section were set first. Each vent has a 2-in. by 6-in. frame that needed a stone sill. The two-piece sills were cut and laid up to the proper height; then, each wood frame was set and leveled on the sill (see photo at right). The sills are sloped slightly away from the house to shed water. Wall ties anchor the frames in the stonework. The tops of the vent frames were flashed with copper. Once the frames were set, we laid stones up to the top of the frames. We built keystone lintels over the flat tops of the vent frames, then turned to the corner stones on the main wall.

The corner stones had been cut to rough dimensions in a quarry in Pennsylvania. They now needed to be hand-dressed and fitted into position. The common stonework in the house was roughly dressed (putting a flat face on stone) with a stone or brick hammer to match the stones on Burgess Lea, but the corner stones required a large face that needed to be brought into a single plane. This was accomplished with a 3-lb. hammer, a heavy point, and a chisel. The 8-in.-thick stones were 28 in. to 36 in. long and up to 14 in. high.

# Laying Wall Stones up to Reference Lines

After putting down a bed of mortar, we set the stones into it. As each stone was adjusted into position, I sighted between the two lines running horizontally across the wall. Because I could move the two horizontal reference lines up and down, I got an accurate sense of where the plane of the wall should fall. The farthest protrusion of stone came to within ⅛ in. of the line.

Venting the crawlspace. To provide ventilation for the crawlspace beneath the house, wooden vent covers are set on sloped stone sills and set into the stone-veneer wall. Galvanized wall ties connect the vent frames to the stone, and copper flashing covers the tops of the frames.

I eased each stone down in the mortar bed until it touched the stone below it. With the mortar filling any voids, the stone is not likely to settle further as weight is added to it. All voids behind the stones should be filled with spalls of stone and mortar to prevent settling. In this manner, the amount of stonework laid in a day is limited not by the curing of the mortar but only by the stonemason's energy and ability.

At the end of each day, excess mortar was raked from between the stones to a depth of 1 in. By that time, the mortar had begun to set up and it fell away cleanly from the stonework. These 1-in.-deep joints would be filled later when we pointed the walls.

*The amount of stonework laid in a day is limited not by the curing of the mortar but only by the stonemason's energy and ability.*

## TIP

*Before pointing, remove heavy mortar buildup on the stonework by washing it down with a muriatic-acid mixture.*

**Keystone lintels over windows.** Wedge-shaped stones, patterned after those on a house in Pennsylvania, provide support over windows. Although flat rather than curved, the wedge shape prevents the stones from sagging into the window frame once the lintel is complete.

## Plan for Windows

It's important to plan for windows, doors, rooflines, and other architectural features so that the stone veneer will achieve a uniform flow. For example, it's distracting to see all 10-in. high corners and then a 4-in. high piece that is used to hit the top of a window or the top of a roofline. Over windows, stones should be used to break the vertical lines that the window frames create.

The tops of the window frames were flashed with copper in preparation for the lintels. On this project, keystone lintels were used (see photo above), eliminating the need for angle iron. However, if angle iron is used, I suggest that it be recessed at least 1 in. behind the front of the window frame and pointed over. Even painted iron will rust, and it's better off hidden from view.

As we headed up between the upper windows, we made a niche for the date stone. Date stones are traditional on period stone houses in Pennsylvania. Learning to carve them has been a natural extension of my work and has proved to be valuable. In fact, I had carved the date stone for Burgess Lea in 1985. Because I had a lot of free time in the evenings in Mississippi, I carved this date stone while the house was being built (see photo at left on facing page). The design of the date stone originated in Pennsylvania. It means good luck, and the symbol appears throughout the Williamses' house. Because the date stone wasn't finished when we were ready for it, we built a plywood dummy and set it temporarily in the niche we created to hold a place for the stone. After the arch stones were set, the dummy was removed, and the real date stone was inserted.

The stonework was then brought up to the roofline. We tucked the top stones behind the trim and forced mortar up on top of the last course of stone and down between the stone and the trim. This created a solid top course and provided a backing for pointing.

## Point Stone after Mortar Is Well Set

Once the stone had been laid up to the roofline and was washed down with water, we were ready for pointing. (Had there been a heavy mortar buildup on the stonework, I would have washed it down with a muriatic-acid mixture.) Pointing weather-proofs and protects the less-stable building mortar within.

**A project for nights away from home.** Faced with long stretches of time far from home, the author detailed this date stone by hand in the evenings. The stone was set in a niche in the middle of the front wall of the house near the eaves.

**Pointing keeps water out.** Mortar is scraped out at the end of each day to a depth of 1 in. After the mortar has set completely, the entire job is pointed. Because of the mix used and the cold joint between the two layers of mortar, the building will not be difficult to repoint.

It's important to stress that building and pointing are two separate processes. Pointing is much like adding a final coat of stucco to the brown coat. The building mortar should be well set before it is pointed. Years down the road, the stonework will need to be repointed, and if the process was handled correctly at the start, it should be simple to remove the original pointing without damaging the stonework or the underlying mortar. I have had the unpleasant experience of trying to chop out mortar in work that the mason pointed at the end of each day. Because of the liberal use of portland cement in most masonry work today, difficult repointing is not an experience I wish to pass on to future stonemasons.

I used a rich lime–sand mixture, which resulted in a bright white, slightly raised pointing. The mixture consisted of 6 parts of fine sand, 1 part white portland cement, and 2 parts lime. You must keep the proportions consistent to maintain uniform color in the pointing. The mortar is mixed to the consistency of soft cream cheese and applied with a ¾-in.-wide trowel.

---

*Steven Snyder is a stonemason and sculptor in Point Pleasant, PA. His sculpture, home, and studio have been featured on HGTV and in* Country Living Magazine. *His Website is www.stonesculpture.net.*

# Vinyl Siding Done Right

■ BY MIKE GUERTIN

I cringed the first time a custom-home client asked for vinyl siding on a new $400,000 house. But I had little choice. Our market was dead. So I took the project. Now more than half of my custom-home clients ask for vinyl, and I actually like installing it. Although vinyl's installation is faster, making vinyl look good on a house is more challenging than wood or fiber cement.

Hands on is always the best way to learn, but you can get basic installation instructions from the Vinyl Siding Institute, and manufacturers offer brand-specific installation manuals as well. Here, I highlight a few techniques for enhancing the appearance of clapboard-style vinyl siding in new construction.

A word of caution: Vinyl siding leaks a lot. Water enters where vinyl panels overlap and at the end gaps inside the trim's receiving channels. Manufacturers even punch weep holes into the bottom edge of siding to let the water escape. Besides being careful to maintain the integrity of the tar paper or housewrap behind vinyl, I recommend flashing windows and doors to direct water back out when it gets past the flanges or casing.

## Begin with the Trim

Vinyl siding depends on a few special trim pieces that either hold everything together or hide the edges and ends of the panels. Undersill trim, for instance, is installed under windowsills or other horizontal projections and grasps the top edge of the siding panel below it (see top right photo on p. 79). The J-channel receives the ends of siding panels where they abut windows, doors, or other walls. Most pieces of vinyl trim, though, go by the same names as their wood-siding counterparts: soffits, fascia, and corner boards.

I start by installing the high trim (rakes, soffits, and fascias). For rakes, I usually bend aluminum coil stock on a sheet-metal brake to wrap the face and bottom edge of a 1x rake board. Manufacturers also offer

**Pushing up locks panels together.** The bottom edge, or butt, of one panel clicks into the locking channel at the top of the preceding course, and the siding goes up.

## Sources

**Vinyl Siding Institute**
(888) 367-8741
www.vinylsiding.org

vinyl-coated aluminum and all-vinyl trim stock that are bendable and give excellent results. The rake board is furred out with a board narrower than the rake to create a space for the siding ends—sort of a site-made J-channel. Prep for soffit installation begins as I frame the house. Along the tails of the rafters or trusses, I install a 2x subfascia, then hang the soffit panels and fascia on that (see photo below). Where soffit panels meet the wall of the house, they are supported by J-channel, undersill trim, or window casing.

## Snapping a Line Keeps Corner Boards Straight

After the rakes and soffits are finished, I install the corner boards. Because corner boards are hollow and flexible, nailing them up straight can be difficult. Snapping chalklines on both sides of the corner ensures that the corner board goes on straight (see photos on p. 76). I use short corner-board sections as templates to mark the wall for the chalkline.

I begin nailing the corner board at the top and use sheet-metal snips to cut the flange back where it abuts the already installed soffit trim. The top two nails (one on each side of the corner) are placed at the top edge of the nail slots and are set tightly. These nails lock the corner-board top in place so that it expands downward rather than upward into the soffit. I drive the rest of the nails (spaced about 12 in. apart) in the middle of the slots to permit that expansion, leaving the head about 1/16 in. proud of the nailing flange so that the vinyl can move.

Site-bent aluminum fascia captures front of soffit panels.

Window and door casing supports soffit panels at wall.

The author cuts soffit panels ¼ in. shorter than the soffit depth to allow for thermal expansion; he then traps the soffit panels with site-bent aluminum fascia trim and vinyl window casing. The window casing captures the top of the siding panels and mimics a frieze board, which looks better than narrow J-channel.

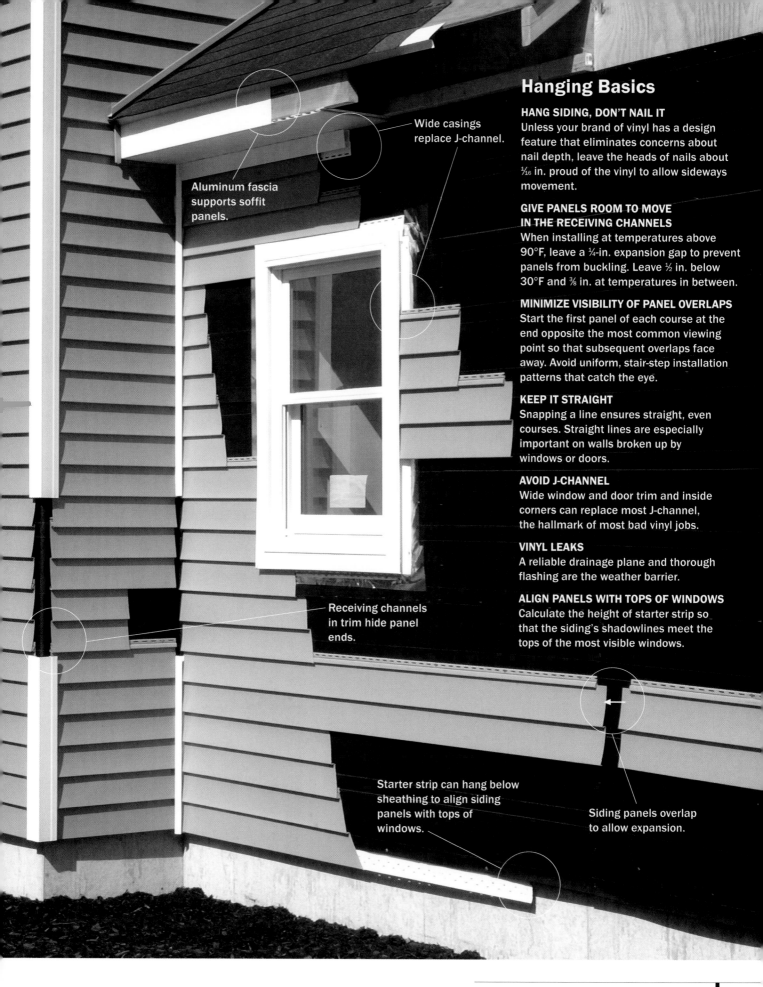

Wide casings replace J-channel.

Aluminum fascia supports soffit panels.

## Hanging Basics

### HANG SIDING, DON'T NAIL IT
Unless your brand of vinyl has a design feature that eliminates concerns about nail depth, leave the heads of nails about $\frac{1}{16}$ in. proud of the vinyl to allow sideways movement.

### GIVE PANELS ROOM TO MOVE IN THE RECEIVING CHANNELS
When installing at temperatures above 90°F, leave a $\frac{1}{4}$-in. expansion gap to prevent panels from buckling. Leave $\frac{1}{2}$ in. below 30°F and $\frac{3}{8}$ in. at temperatures in between.

### MINIMIZE VISIBILITY OF PANEL OVERLAPS
Start the first panel of each course at the end opposite the most common viewing point so that subsequent overlaps face away. Avoid uniform, stair-step installation patterns that catch the eye.

### KEEP IT STRAIGHT
Snapping a line ensures straight, even courses. Straight lines are especially important on walls broken up by windows or doors.

### AVOID J-CHANNEL
Wide window and door trim and inside corners can replace most J-channel, the hallmark of most bad vinyl jobs.

### VINYL LEAKS
A reliable drainage plane and thorough flashing are the weather barrier.

### ALIGN PANELS WITH TOPS OF WINDOWS
Calculate the height of starter strip so that the siding's shadowlines meet the tops of the most visible windows.

Receiving channels in trim hide panel ends.

Starter strip can hang below sheathing to align siding panels with tops of windows.

Siding panels overlap to allow expansion.

# Alignment Is Crucial

Adjusting the starter-strip height can keep panels even with the tops of windows and doors. Here, the author dropped the starter strip below the sheathing by pop-riveting it to pieces of coil stock.

Chalklines keep corner boards straight.

Pop rivets hang starter strip from coil stock.

Chalklines keep panels straight and align them with tops of windows.

Snapping chalklines for corner boards and siding panels ensures that the flexible vinyl stays straight. Siding panels have some vertical play, so checking overall siding height frequently allows installers to make small adjustments and avoid accumulated error.

After determining exactly where at the bottom of the wall the starter course will begin (see top right photo on facing page), I trim all the corner boards just below the bottom of the starter strip (¼ in. in winter, ½ in. in summer) and finish nailing them in place. Except for window trim, most of the prep work is now done.

# Accommodating Thermal Expansion

Vinyl-siding manufacturers try hard to make their panels look like real wood. Although the manufacturers are getting better, vinyl installs, behaves, and performs much differently from the natural material it mimics.

## Nailing Details and Expansion Gaps Accommodate Panel Movement

Leave nail heads ⅟₁₆ in. high to allow movement.

Tin snips notch field-cut panels.

Where panels overlap, factory-notched nailing hems keep adjacent panels from buckling. New notches must be made where panels have been cut.

Face overlaps away from most common viewing point to minimize their visibility.

Panels overlap so that they can slide past each other as they expand and contract with temperature changes. To allow this movement, the panels are made with the nailing hems notched at the end. This notch must be duplicated on field-cut panels. Most nail heads are left high to avoid trapping the panels. After the hems are locked, the endlap is backed off for no more than 1 in. overlap at the ends of the panels.

Butting panels too tightly into trim's receiver channels can cause buckling in high temperatures.

Fixing the center of a panel with two nails ensures that thermal expansion happens equally at each end.

Whereas wood moves in response to moisture content, vinyl moves in response to temperature changes. And it can move a lot through expansion and contraction, especially along the length of the clapboard-style panels. To compensate for this movement, adjacent siding panels are overlapped so that they can slide sideways past each other without buckling or creating gaps. Nails are left proud of the surface to allow for this movement. Where panels meet windows, doors, or corner boards, they are cut short. Receiving channels on the trim conceal the expansion gap. Trim along roof slopes should be set ½ in. above hot asphalt shingles.

## Minimize J-channel around Openings

Windows and doors need to be surrounded with a receiver channel to accept the ends of the siding. This process usually means wrapping them all with J-channel, one of the dead giveaways that a house is covered in vinyl. Minimizing J-channel takes planning and creativity. When I know a house will be sided in vinyl, I select windows that have either an integral J-channel or a slot to receive an applied cap that creates a receiver channel. Alternatively, I use vinyl window and door casing, which is essentially a 2½-in. wide J-channel (see photo on p. 75). The greater width can enhance the window rather than detract from it, as a narrow band of standard J-channel does.

Around doors, I use a couple of other tricks to build a receiver channel. Using either a preformed foam plastic or a custom wood door surround is the easiest. I simply apply a 1x3 wood spacer over the door jamb and exterior sheathing (see photo at right on p. 80). Then I fasten the decorative trim over the spacer to create a ¾-in.-wide pocket that receives the ends of the siding. Alternatively, I can trim the door with furring and bend aluminum coil stock around it.

**Notching siding into J-channel keeps water outside the wall.**

When my bag of tricks runs out and I'm forced to wrap an opening with J-channel, I try to make it look as clean as possible. You can't count on the leg of the J-channel at the top of a window or door to serve as head flashing. When using J-channel, I always install a drip cap that slips beneath the housewrap and laps over the window or door. The J-channel goes over the flashing.

## Align Siding Panels with Window Tops and Intersecting Roofs

Perhaps the most fundamental trick to making vinyl look good begins well before installation. One of my goals when planning wood siding is to have the bottom edges of the clapboards line up with the tops and bottoms of windows and doors whenever possible. This alignment unites the exterior look of the building (see "Installing Wood Clapboards," p. 81). Alignment is difficult with vinyl because the course exposure isn't adjustable. And this lack of alignment is one of vinyl's telltale signs.

To minimize this problem, I begin planning the vinyl course layout before the foundation of the house is poured. Because the limiting factor is the vinyl itself, I choose the siding for the house first. The three most common vinyl-siding patterns are double four, triple three, and double five. That is, vinyl panels are made to look like two 4-in. clapboards, three 3-in. clapboards or two 5-in. clapboards. Their total exposures are fixed: 8 in., 9 in., and 10 in.

Knowing the exposure allows me to plan at least some of the rest of the house around it. I plan foundation drops, those spots where the foundation steps up or down to keep pace with the grade, on increments to match the siding height. It's easy to position the window rough openings so that the tops of windows match a siding course. The bot-

# Window Trim Is Not Flashing

Whether it's J-channel or a wider piece of casing, a tab cut in the top piece of the window trim wraps over the side to divert water. In addition, the metal head flashing tucks behind the tar paper and over the window top or flashing tape is used to seal tar paper to the window's nailing fin.

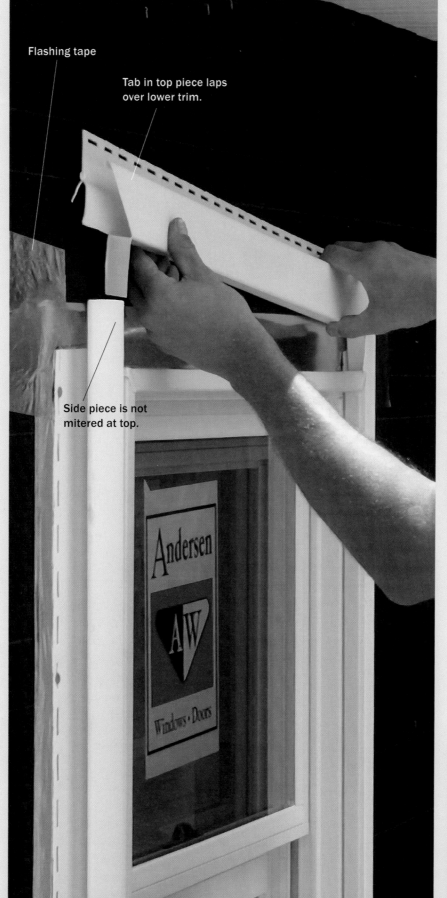

Flashing tape

Tab in top piece laps over lower trim.

Side piece is not mitered at top.

Undersill trim captures top of cut panel as it wraps around window.

Tar paper carries water shed by flashing past nailing hem of next panel.

**Behind the window casing,** the undersill trim secures the cut vinyl panel under a window. Builder's felt and flashing membrane direct water past the nailing hem of the first full panel below the window, pushing the water outside.

When notching panel at bottom of window, leave ¼-in. expansion space all around.

toms of windows are hit or miss unless I'm using custom-built vinyl windows and can specify window height.

When I have to notch panels at the bottom of windows and at J-boxes (see bottom left photo), I make sure the notch isn't at a joint between panels—both because it's easier to cut one notch than two and because it looks better. I leave a ¼-in. space on the rip beneath the opening and on each side of the window or box to allow for expansion space. I can't adjust the height of the doors, but the width of the surrounding trim can be tweaked to align with the bottom of a siding panel if it's within 1 in. or so.

When porch, garage, or other intersecting roofs interrupt siding, I have to install J-channel to start the panels. But by adjusting the roof pitch just a little so that the butt line matches the roof–wall intersection, I get a clean look. The same goes for the height of a deck. It may seem excessive to plan exterior building elevations to match the siding, but the results are worth the effort.

*Mike Guertin is a contributing editor to* Fine Homebuilding *magazine and the author of* Precision Framing *and* Asphalt Roofing, *both published by The Taunton Press. He lives in East Greenwich, RI.*

## J-boxes and Furred-Out Door Trim Replace J-Channel

Trimming around outlets, light fixtures, electric-meter sockets, and ventilation exhaust ports can be accomplished easily with preformed J-boxes. Several manufacturers make special accessory boxes with integral J-channels or snap-on J-surrounds to make trimming the penetration faster, more weather-resistant, and more attractive than something that can be fabricated on site. Whenever possible, I roughly locate the J-boxes on the wall but don't attach them until the siding goes up. This way, I can achieve a cleaner look by precisely locating the top of each box against the bottom of the panel above it.

Around doors, the easiest way to hide panel ends is to fur out the trim so that siding panels abut the furring and the trim forms a receiving channel.

Furring lifts door trim over panel ends.

**Utility boxes create receiving channels.** A snap-on molding holds panel ends around a light box near the front door.

**Door trim replaces J-channel.** At a front door, a furring strip lifts the door trim enough to let siding panels tuck in behind.

# Installing Wood Clapboards

■ BY RICK ARNOLD AND MIKE GUERTIN

**W**e've installed clapboards just about every way but upside down. We've tried different nailing locations and patterns and used various hand-driven and pneumatic fasteners. We've installed clapboards from the bottom up and from the top down, and we've organized our installation teams several ways. Although we wouldn't call any of our experiments failures, we have found a layout method, fastener and fastening pattern, crew size, and details that ensure good performance and efficient installation.

> *Good layout results in a look that sets clapboards apart from vinyl siding, whose layout shows no regard for window and door height.*

# A Few Details before Starting

Before we lay out and install clapboards, several details need to be in place. We have up whatever housewrap we're using, and install the corner boards (see drawings on facing page). Corner boards provide a stop for the clapboards so that they don't have to be mitered or coped. We make our inside corners out of 1x2 or square 5/4 stock. Anything less would be too thin and leave the painter no room to caulk between the inside corner and the ends of the clapboards.

Outside corner boards avoid the need for delicate mitered joints at locations easily damaged by lawnmowers and other equipment. We make ours from 1-in. stock and prime any raw edges before installing the siding. Doing so minimizes shrinking and swelling and just about eliminates stains from dissolved tannins leaking to the face.

For a heavier look, we sometimes build outside corners from 1x3 furring and apply 1x6 finish corner boards over them after the siding goes on. We nail the outside edge of the finish corner board to the clapboards every 16 in. to 24 in., being sure to nail through the thick bottom of the clapboards. The clapboards support the overhanging corner board, and nailing is less likely to crack it. Overlapping clapboards with the corner board in this way is a more watertight detail, but the nooks where the clapboards go under the corners can provide homes for insects.

When we apply the corner boards, we let them run long. Then we trim them to length once the water table—the horizontal board where the wall meets the foundation—has been installed (see photo on p. 81).

On gable ends without overhangs, we install 1x6 rake boards over 1x3 furring that runs flush with the top of the roof sheathing. We leave enough room to slide the clapboards at least 3 in. under the 1x6 (see bottom drawing on facing page). On overhanging rakes, we cut the clapboards to the rake angle and butt them to the soffit. Then we trim this joint with a piece of 1x3.

# Careful Layout Is Crucial to a Good Job

The common clapboard in our area is ⅜ in. thick at the butt end and 5¼ in. wide. Clapboards typically have 4 in. of exposure to the weather, but we tweak the layout so that the bottom of one course lands directly on top of the windows and the bottom of another course matches the bottoms of the windows (see photo on p. 89). We also want no more than a ⅛-in. variation between adjacent courses. With luck, we can make the bottom of a course fall directly on top of the doors, too. Good layout results in a look

## The Common Clapboard

The authors nail about 1 in. from the bottom, where the clapboard is supported by the lower course.

5¼ in.

⅛ in.

Felt spline

4-in. to 4½-in. exposure

⅜ in.

## OUTSIDE CORNERS SERVE TWO PURPOSES.

They are a way to turn a corner without using a compound miter, and they protect the clapboards from damage at this exposed location. Overlapping 1x4s form the simplest corner. For a heavier look, nail on a subcorner of 1x3 furring, butt the clapboards to the 1x3, and apply a finish corner made of 1x6 that laps the clapboards and the water table.

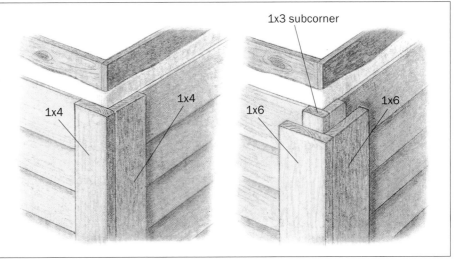

## INSIDE CORNER BOARDS MAKE COPING OR MITERING THE CLAPBOARDS UNNECESSARY.

Butt the clapboards to a vertical piece of 5/4 square stock or 1x2 at the inside corners. Both options leave plenty of room for a neat caulking job.

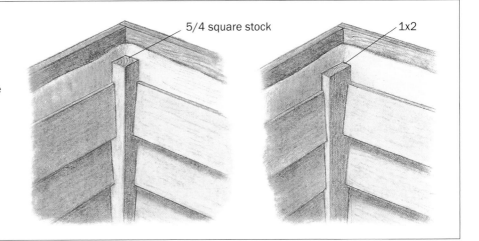

## A NEAT FINISH WHERE CLAPBOARDS MEET THE GABLE.

On walls with no overhanging gable, nail 1x3 furring to the house and go over this furring with a 1x6 finish rake board. The tops of both are flush with the top of the roof sheathing. The roof shingles extend over these boards, and the clapboards tuck into the space behind the 1x6. On houses with an overhang, cut the clapboards to the rake angle, butt them to the soffit, and cover this joint with a trim piece.

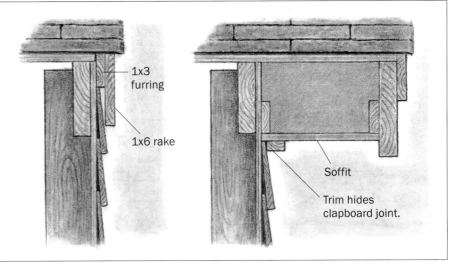

## Layout with a Story Pole

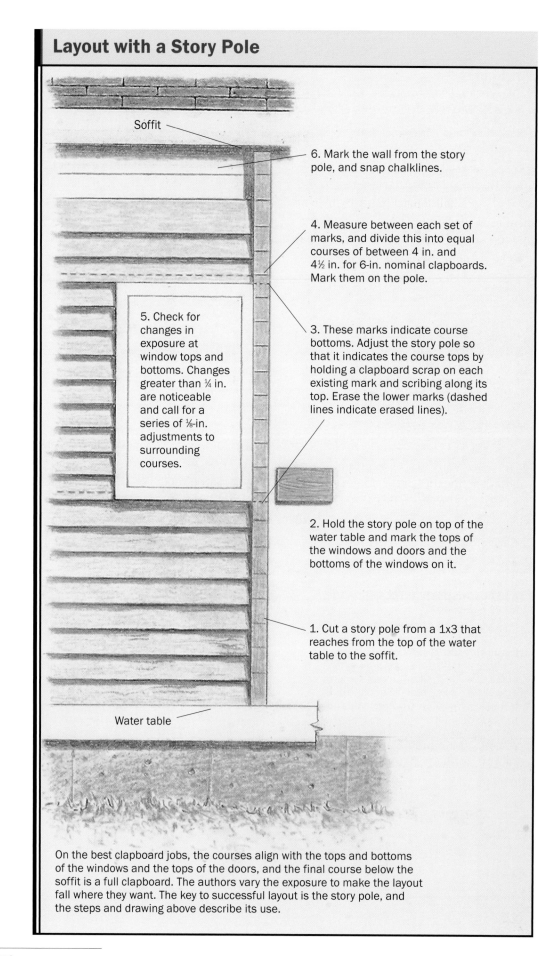

Soffit

6. Mark the wall from the story pole, and snap chalklines.

4. Measure between each set of marks, and divide this into equal courses of between 4 in. and 4½ in. for 6-in. nominal clapboards. Mark them on the pole.

5. Check for changes in exposure at window tops and bottoms. Changes greater than ¼ in. are noticeable and call for a series of ⅛-in. adjustments to surrounding courses.

3. These marks indicate course bottoms. Adjust the story pole so that it indicates the course tops by holding a clapboard scrap on each existing mark and scribing along its top. Erase the lower marks (dashed lines indicate erased lines).

2. Hold the story pole on top of the water table and mark the tops of the windows and doors and the bottoms of the windows on it.

1. Cut a story pole from a 1x3 that reaches from the top of the water table to the soffit.

Water table

On the best clapboard jobs, the courses align with the tops and bottoms of the windows and the tops of the doors, and the final course below the soffit is a full clapboard. The authors vary the exposure to make the layout fall where they want. The key to successful layout is the story pole, and the steps and drawing above describe its use.

that sets clapboards apart from vinyl siding, whose layout shows no regard for window and door height.

By the way, layout is quick and easy if you set all the windows at the same height. If their height varies, plan the differences between windows to match clapboard-course increments.

Make sure the top of the water table, usually 1x6 stock, falls on a 4-in. to 4½-in. increment from the bottom of the windows and that its bottom overhangs the foundation by 1 in. If the foundation steps down as the grade changes, we lower the water table by course increments.

The story pole is the crucial layout tool. We cut it from 1x3 furring and make it long enough to reach from the top of the water table to just beneath the soffit. On this job, we needed two story poles because a second-floor overhang made it impossible to use one that was two stories high.

With the story-pole bottom atop the water table, we mark the tops and bottoms of each window and the tops of the doors on the pole. We indicate which mark belongs to which window or door to avoid confusion. These window and door locations are the primary marks.

The tops of the clapboards have to be marked on the wall. If the bottoms were marked, each clapboard would cover up the mark for the next one. The primary marks indicate clapboard bottoms. To make the adjustment, we hold a piece of clapboard to each mark and to the bottom of the pole and draw a line at its top. We then erase or cross out the original marks. Now we have primary marks indicating the tops of clapboard courses that match windows and doors.

Starting at the bottom of the story pole, we measure to the next higher mark and divide this distance into equal courses of 4 in. to 4½ in. We do the same thing between each pair of primary marks. Sometimes the space between primary marks is too awkward to divide, say 10 in. There's no way to match a 4-in. course to that space. Usually, it's on the back or side of the house with a shorter bathroom or kitchen window or it's at the top of a door, so it doesn't concern us. We erase the out-of-sync primary mark to avoid confusion, knowing that we'll be notching a clapboard to fit around a window that is off-layout. After we divide the sections, we check for abrupt changes in exposure; those greater than ¼ in. are noticeable. The biggest differences occur just after a primary mark.

Because a primary mark indicates a door or window, we can't change it. We blend the change over a couple of the surrounding courses, making it less obvious. For example, say the courses change from 4 in. to 4⅜ in. We would make one of the 4⅜-in. courses 4⅛ in., then a couple courses at 4¼ in. and a couple at 4⅜ in. To make up the difference, we mark a few at 4½ in. and go back to 4⅜ in. Once we select the final course marks, we square them off and erase any errant marks.

With the pole complete, we use it to mark the sides of all corner boards, windows, and doors. We mark and snap all the lines within reach and begin installing clapboards. As we raise the staging, we mark and snap the rest.

One carpenter nailed on corners and the water table, another cut the starter clapboards. He planned the cuts so that the joints are randomly staggered. Efficient crew members leapfrog each other, cutting or measuring ahead so that there is no downtime.

## The Water Table Is the Starting Point

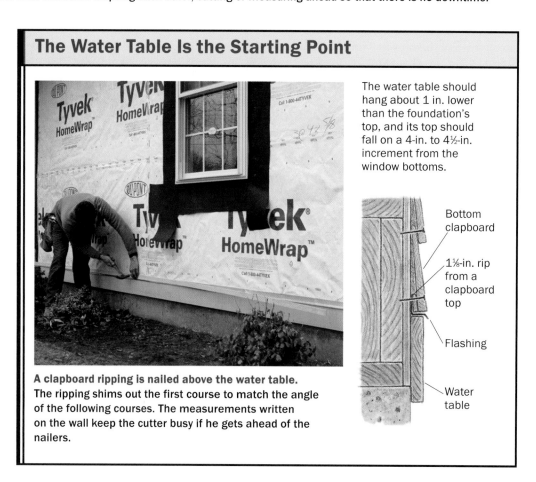

A clapboard ripping is nailed above the water table. The ripping shims out the first course to match the angle of the following courses. The measurements written on the wall keep the cutter busy if he gets ahead of the nailers.

The water table should hang about 1 in. lower than the foundation's top, and its top should fall on a 4-in. to 4½-in. increment from the window bottoms.

Bottom clapboard

1⅛-in. rip from a clapboard top

Flashing

Water table

## Three Carpenters Make an Efficient Crew

One carpenter cutting and two nailing keep the momentum going the best. We first install the water table and cap it with an aluminum drip cap that extends about 1½ in. up the wall and hangs over the water table by ¼ in. or so. Then we measure short spans between windows and count how many courses high the windows are. We write these numbers in crayon on the wall. The cutter can spend his downtime precutting those sections. The cutter begins by squaring the factory ends and setting up starter packs of staggered lengths for long sections of wall. We insist on random staggers. Nothing's worse than seeing equal 1-ft. staggers forming a pattern that catches your eye.

We nail a 1⅛-in. ripping from the top of a clapboard to the wall above the water table (see sidebar on facing page). This shims the bottom of the first clapboard out to match the succeeding courses. The measurement 1⅛ in. is the common overlap of 4-in. exposures of 5¼ in. (finished size) clapboard, minus the saw kerf. We use the leftover 4-in. bottom rip as the last piece under the soffit.

The nailers tack up the starter clapboards and measure the subsequent pieces before nailing off. The cutter can ready pieces with no downtime. Installers often tack the clapboards on a whole face and finish nail when they crank the staging down. By keeping ahead of each other, by precutting or by premeasuring, the cutter and nailers always have something to do.

**All raw wood is primed.** Water entering through end grain gets under the surface and causes paint to peel. Back-priming and end-sealing minimize this problem.

*Even when we've sanded, we've had paint fail on smooth clapboards.*

## Details for Appearance and Weather Resistance

There are several reliable details we use on every job. Although many siders bevel-cut their joints, we butt ours. We have no shrinkage problems with the dry finger-joint primed material we use, and square cuts look better over time. We keep a pail of primer handy and dab cut ends to seal out water. We insert tar-paper or aluminum step-flashing cards behind each seam. They lap the lower clapboard by ½ in. or so and shed any water that penetrates the joint. (see top photo on p. 88). We keep the clapboards ⅟₁₆ in. to

Tar-paper splines ensure a leakproof joint. Placed under each butt joint and at the windows, the splines lap the lower course by ½ in. and guide any water that enters the joint to the outside of the clapboards.

Tar-paper splines lap each other from top to bottom. Finally, they lap the top of a lower clapboard. Any water that gets behind the siding is shed back to the outside.

⅛ in. back from windows. We've had clapboards swell and pinch the windows so that the sash wouldn't move.

A bead of paintable silicon caulk around the window seals out water and lets the clapboards move without distorting the jamb. In high-wind areas where leakage from driven rain is a concern, we caulk the clapboard ends before butting them to the corner boards or casings.

We put tar-paper or bituminous-membrane splines around window flanges for protection from water. The head spline laps the side splines, which lap the bottom spline. We lap the bottom spline ¼ in. to ½ in. over the top of a lower course of clapboard (see photo at left).

Where a wall, such as a dormer sidewall, meets a roof, we lay a clapboard flat on the shingles with its thin edge against the wall to act as a spacer. We then cut, prime, and nail the clapboards to the sidewall, tight against the spacer. When we remove the spacer, there is a clean ¼-in. space between the siding and the roof shingles. Without this space, the siding could wick up water, blistering paint and hastening decay.

When we get to the top of a window, the course just below butts the casing on two sides. The tops of these clapboards extend 1¼ in. higher than the window, but there is nothing above the window. To shim out the next course, we rip the top of a piece of clapboard to 1⅛ in., just as we did above the water table. We cut it to fit between the projecting tops of the lower course, nail it in place above the window and move to the next course (see photo below).

Clapboards come with one side rough and one side smooth. We guarantee finishes only on rough-side-out installations. Smooth-side-out installations require labor-intensive hand-sanding to break the shiny surface left by the planing process or the primers won't bond. We've found that the smooth side of preprimed clapboards needs a light sanding before finish-painting. Even when we've sanded, we've had paint fail on smooth clapboards.

**The top of a clapboard fills the space above the windows.** This keeps the next course from bowing inward as it's nailed. With vinyl windows such as these, the authors leave a ¹⁄₁₆-in. to ⅛-in. expansion gap between the windows and the clapboards and caulk it.

# Nailing Wrong Means More Than Hitting Your Thumb

Fastener location and pattern are controversial issues. We'll admit it: We nail in the wrong place. We nail too low; everybody we know nails too low, and they know it, too. You're supposed to nail through the lower part of the clapboard just above the top of the clapboard beneath. In theory, this gives the boards freedom to expand and to contract in height. If you nail too low, as we do, you'll trap the clapboard with nails high and low, and the clapboards will split as they move. In theory.

The reality is that we've never had a problem when we nail too low, say within 1 in. of the bottom. But we have had problems when we nail "properly." The upper board splits at that point because it has no support from below. Sure, we could nail loosely to try to avoid splitting, but what about the customers who want the nails set and the holes filled? Splits galore. We're not the only ones who nail wrong. We've remodeled homes built in just about every decade of the past 250 years and found similar "poor" workmanship in most. Nuff said.

The proper nailing pattern for clapboards is just as controversial. Should you nail randomly, or should you nail into the studs? The only time we purposely nail into the studs is when a house is sheathed with foam board or when the customer thinks it's a quality issue. We sheathe most of our homes with $\frac{7}{16}$-in. oriented strand board (OSB). OSB holds nails well, so we don't have to hit the studs.

Nailing into the studs causes aesthetic defects. First, the uniform vertical lines of nails belie the horizontal nature of clapboard. Second, the clapboards are drawn more tightly to the wall at the nails. Nailing into the studs creates noticeable waves in the siding. And a 2-ft. stud spacing makes the problem worse. The only solution is random nailing. When we argue the issue with other builders, they insist that we're cheating, that random nailing is easier. Actually, it's not. Easy is having marked vertical lines to nail on. It's difficult to be carefully random and to maintain regular 12-in. to 18-in. nail spacing. We take great care not to have one nail atop another. Random nailing works well for us because we locate our fasteners thoughtfully.

# We Finally Found the Right Fasteners

We have tried several fasteners over the years; some worked, some didn't. Pneumatic staples driven parallel with the grain leave tiny slots that need to be filled before painting. The galvanized staples hold well, and the gun countersinks them well. But we rarely use this method because we install the clapboards rough side out, and the smooth, puttied slots stand out. On smooth-side-out installations, we might consider staples.

We've tried pneumatically driven galvanized ring-shank box nails. We could adjust the gun so that the nails would either be set or be flush. Nailing was fast. But the electro-galvanized finish wasn't thick enough, and they'd rust after a year.

Hot-dipped galvanized ring-shank shake nails were the old standard; we used them for years. We'd set the nails flush and paint over them. Two problems developed. Hammering the nails wears off the galvanized finish, causing them to rust. The tannins in red cedar and redwood react with the galvanization and cause streaking that goes right through the finish paint or stain.

Our final solution is stainless-steel ring-shank siding nails set flush. They cost five times what galvanized nails do. That sounds like a lot, but it's $10 vs. $50 on a house. That's money well spent. The waffle pattern on the nail head blends in with the rough texture of the siding, so with random nailing, they almost disappear.

## Choosing a Wood Species for Clapboards

We've applied clapboards of several wood species over the years. Red cedar is the most common, and redwood a sometime substitute. We've seen pine and spruce on occasion, and mahogany only rarely. Knotty pine, has a poor record of splitting, warping, and shrinking. We avoid it completely. Radially sawn clear pine might be better, but we haven't used it. Spruce is a newcomer in our region. Although we haven't tried it yet, we've heard good reviews from those who have. Mahogany is seldom available, and we've found it hard to nail without splitting. Otherwise, it performs nicely without shrinking or warping, and it takes finishes well.

Red cedar comes in several grades. Clear vertical grain performs the best and is the most costly. A and better performs well and is what we use the most. Because most of our customers prefer to paint clapboards, we pay a little extra for factory-preprimed stock. Preprimed wood resists absorbing moisture from the back that can cause the paint to blister and peel. It nearly eliminates cupping and extractive bleed. Before preprimed material was readily available, we back-primed the clapboards ourselves to get a better finished product. We don't miss the mess. Some of our first preprimed clapboard projects still sport their original paint and after nearly 13 years look like new.

To be both environmentally and cost conscious, we began installing preprimed, finger-jointed clapboards 12 years ago and never looked back. We no longer have to use five bundles of 4-footers on the back of the garage; finger-jointed clapboards are manufactured in 16-ft. lengths. Occasionally, there's a bad section, but we can easily cut it out before installation. We've had to replace only two or three boards in the last 150,000 lin. ft.

## Real Clapboards vs. the Vinyl Alternative

In our area, ½-in. by 5¼-in. finger-jointed preprimed A and better clapboards cost $160 per square. Good-quality vinyl siding goes for $60 per square. It takes about 50% longer to install clapboards than vinyl siding. Then there's the cost of painting or staining the entire house every 5 years. Even in conservative, slow-to-change New England, we're seeing less clapboard siding going on new homes. Economics and high maintenance are taking their toll.

Still, vinyl siding just doesn't look as good. It fits awkwardly around windows. Seams are obvious when viewed from the wrong direction. There's debris-catching J-channel everywhere. Vinyl creaks from expansion when the sun hits it. The best vinyl siding is said to look almost like clapboards. Almost. But there's nothing like the real thing.

*Rick Arnold and Mike Guertin* are contributing editors to Fine Homebuilding *magazine and the co-authors of* Precision Framing. *In addition, Rick Arnold is the author of* Working with Concrete *and Mike Guertin is the author of* Asphalt Roofing, *all published by The Taunton Press.*

# Working with Fiber-Cement Siding

■ BY JOHN LA TORRE, JR.

**T**he words *fiber-cement siding* may conjure up cold industrial visions, but this siding is nothing like cement. Ever since I started using fiber-cement siding, I've recommended it to anyone who would listen. When painted, it looks every bit as warm and inviting as wood siding. And in my area, at about $1.80 per sq. ft. installed, fiber-cement siding is similar in price to stucco and is actually cheaper than cedar siding.

## Preparation: The Old Rules Still Apply

As with other types of plank siding, I sheathe the house with ½-in. CDX plywood and then install housewrap over the plywood. Next, wood trim is installed around windows and doors and at building corners.

I have seen fiber-cement siding installed with ¾-in.-thick trim, but I prefer the sturdier look of 1½-in.-thick trim. I chose rough-sawn hemlock as the trim stock for this project because its texture closely matches the fiber-cement siding I used. Fiber-cement siding can have a variety of textures (see sidebar on p. 94).

Fiber-cement siding has to be nailed through to the studs (not just to the sheathing), so I draw vertical lines at each stud location on the housewrap with a felt-tipped pen. As with other types of plank siding, I install galvanized drip edge as flashing over each window and door.

The final preparatory step is installing a spacer along the bottom edge of each wall to kick out the bottom plank to the proper angle. Lath or a narrow strip of wood can be used as a spacer, and occasionally, I use long, narrow strips of fiber-cement siding left over from ripping down planks (see top photo on p. 95). My spacer of choice, though, is ⅜-in. galvanized Z-metal (see bottom photo on p. 95), which is just the right thickness and won't rot or swell from rain splashing underneath. I also use spacers over windows and doors if there is a full-height plank above.

**Stealth nailing.** Fiber-cement siding is flexible across its width, so nailing along the top edge draws it in tight along the bottom. Each course hides the nails on the course below.

*These planks are so rigid and uniform that even when held by a single nail in the middle, they stay nice and straight.*

## Many Faces of Fiber Cement

The basic composition of fiber-cement siding is portland cement, sand, and cellulose fiber. This siding contains no asbestos, glass fibers, or formaldehyde. The only construction hazard is the potential inhalation of silica dust. Because the composition is mainly cement and sand, manufacturers guarantee the siding for 50 years against rotting, splitting, warping, buckling, or swelling, and fiber-cement siding is resistant to rain, hail, snow, salt air, and termite and insect damage. Fiber-cement siding is also non-combustible.

All fiber-cement planks are 12 ft. long and ⁵⁄₁₆ in. thick. On this job, I used 8¼-in. siding manufactured by the James Hardie Company, but many other widths are available, including 5¼-in., 6¼-in., 7¼-in., 9¼-in., and 12-in. planks. Each width is supposed to be installed with a 1¼-in. overlap.

Fiber-cement siding also comes in a variety of finishes, including cedar (wood grain), smooth, beaded cedar, and beaded smooth.

(Siding samples: Courtesy of Cemplank)

# Careful Layout Makes the Job Go More Quickly

Layout for fiber-cement siding is the same as for any other type of horizontal siding once you've chosen a plank width and approximate exposure. For this job, I chose 8¼-in. planks with a 7-in. exposure.

I start each wall by putting on the bottom plank. This plank can be cheated down a little if need be to make the rest of the layout come out even. When the bottom plank is installed, I mark the top of each course every 7 in. up the wall. It's rare when my layout leaves me a full plank at the top of the wall as well as at the tops of the windows and doors, but by adjusting the courses just ¼ in., I can gain or lose up to 3½ in. on an 8-ft. wall.

When the layout goes around the corner and onto a wall that drops in grade, I project the lowest course line from the adjacent wall and snap a level chalkline. I then mark courses both up and down from that line, adjusting the layout as needed.

# You Must Leave Space for Caulk

With course marks on the wall, I'm ready to begin installation. A word of caution about handling 12-ft. fiber-cement planks: Don't pick them up in the middle when they're lying flat. They may snap in two, and if they don't break, they can bend so much that they distort and won't lie flat on the wall. Instead, turn the plank on edge before picking it up.

I know companies that throw a small army at a siding job. One or two guys do the cutting while several crew members nail siding to the house. Fiber-cement siding is easy enough to work with that a crew used to

Bumping out the bottom. To get the siding started at the proper angle, a spacer should be installed behind the bottom edge of the bottom plank. A strip ripped from a plank (top) or galvanized Z-metal (bottom) can be used as a spacer.

installing wood siding should have no problem switching over, given the right tools.

I usually have only one other guy on my crew. Rather than having him cut for me and then hang around waiting for my next measurement, I prefer to cut and install the siding by myself, which frees him to prep the other walls while I work behind him.

The only difficulty in working alone is aligning the planks. Here's how I do it. I

**TIP**

*Start each wall by putting on the bottom plank. This plank can be cheated down a little if need be to make the rest of the layout come out even.*

drive nails partially at each layout mark as guides at the ends of the wall (see top photo on p. 96). I then push the plank up to the nails and hold it with one hand while I drive a nail in at the stud line with the other (see bottom photo on p. 96). These planks are so rigid and uniform that even when held by a single nail in the middle, they stay nice and straight.

**Guide nails speed installation.** Course widths are measured up and marked from the top edge of the bottom plank. A temporary nail is driven at each mark as a guide.

**Nails are driven at the studs.** Vertical lines are drawn on the housewrap at every stud location. After guide nails are driven, each plank is pushed up to the guide nails, and a nail is driven through the top edge of the plank at each stud.

The manufacturer, in this case James Hardie, recommends leaving a ⅛-in. gap between the end of the plank and any trim, as well as between butt ends of the planks on walls more than 12 ft. long. The gaps are then filled with caulk, which allows for a small amount of expansion and contraction in the planks. Leaving gaps also means that it's not absolutely necessary to make perfect cuts. I try not to be sloppy, but any slight inaccuracies in my cut are hidden by caulk.

Once the plank is tacked in place, I simply go ahead and drive a nail at each stud. The manufacturer also recommends driving a nail within ½ in. of the ends of the planks, which requires predrilling. The nails on each side of a joint can then both catch a stud. But I've found that as long as the studs are straight and not bowed, joints in the siding can fall midway between studs and still be held tight with nails in the studs adjacent to the seam. On a long wall, this feature lets me use full 12-ft. planks that don't need cutting to land on studs and don't need predrilling for nails on the butt seams.

# Blind-Nailing with Your Eyes Open

As each plank overlaps the one below, a small triangle of empty space is created behind it. Fiber-cement planks are rigid along the length but slightly flexible across their width. When a nail is driven 1 in. down from the top edge, this flexibility allows the plank to be sucked in tight to the one below, deflecting the plank into a slightly inward curve (see photo on p. 93).

With the planks held in place this way, few if any fasteners are needed along the bottom edge, where they would be visible. This technique, called blind-nailing, means that there are no nails to set and caulk, resulting in a cleaner finished look after the siding is painted. However, blind-nailing is recommended only on planks less than

(see photo on p. 93).

*Overdriving nails causes siding to shatter under the nail head, which weakens the siding.*

Fragile corners; nail with care. Two places where predrilling is necessary are plank corners near trim boards (above) and long angled points where planks are cut to follow the angle of the grade or the roof (below).

Catching the stud. When the end of the plank lands over a stud, the author predrills a hole to get the nail as close to the corner as possible.

**C**utting siding made from sand and cement requires a few special tools. Most cutting can be done dust-free with an electric hand shear (see photo at right). A shear can't cut into corners, so a jigsaw with a carbide blade finishes the cut (see photo at left on facing page). A carbide hole saw can be used to cut in a round electrical box (see middle photo on facing page), and a small circular saw with a carbide blade can be plunged in to cut for a square box (see photo at right on facing page). A dust mask must be worn with the last three cutting choices.

**Electric hand shear means no dust.**

12 in. wide and only when stud spacing is 16 in. or less.

The manufacturer recommends a corrosion-resistant siding nail. I use 2-in. galvanized roofing nails because their heads have such a flat profile. A pneumatic nailer can be used to attach the siding, but I don't use one because it can be difficult to regulate nail depth. In fiber-cement siding, nails are supposed to be driven flush. Overdriving nails causes siding to shatter under the nail head, which weakens the siding.

One thing that surprised me about installing fiber-cement siding is that predrilling nail holes is rarely necessary. Roofing nails sail right through the stuff, even easier than through hardwood. The only places you must predrill are at the ends of the plank where a nail close to a corner could cause the corner to break off (see top right photo on p. 97 ) or where there is little siding to nail

through, such as a piece tapered to an angle (see bottom right photo on p. 97).

I use a carbide-tipped masonry bit for predrilling, but it is not necessary to drill all the way through the plank. Drilling just halfway through is enough to prevent breakage.

## Cutting Sand and Cement

As you'd expect, some specialized tools are required for cutting fiber-cement siding. A circular saw equipped with a carbide-tipped blade will cut the planks. But cutting in this manner is not recommended because the fine silica dust produced is a potential carcinogen. In fact, each plank carries a warning label describing the risk of silicosis, a disabling lung disease. Makita has recently come out with a hand-held 7¼-in. circular

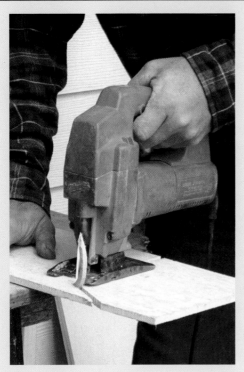

Jigsaw gets into corners.

Hole saw is used for round holes.

Plunge-cutting makes square holes.

saw specifically designed for cutting fiber-cement siding; it's equipped with a dust-collector housing. This saw can cut up to five pieces of stacked siding in one pass. But proper use of this saw requires that it be connected to a shop vac with a good dust-inhibiting filter.

Fiber-cement planks can also be cut with a knife using the score-and-snap method similar to drywall, but it is almost impossible to get a clean cut this way. The best, least-dusty way to cut fiber-cement siding is with a hand-held electric shear (see photo on facing page). And the electric shear works so quickly that I use it even for ripping 12-ft. planks.

A shear works on the same principle as a sheet-metal nibbler, but the jaws are larger. The shear cuts a segment or kerf about ¼ in. wide and makes virtually no dust. My local

supply store rents these shears, and they'll lend a shear free of charge if I buy siding there. After I'd rented a shear a couple of times, I decided that I wanted my own and bought a Snapper™ for about $200.

When cutting with the hand shear, I try to make cuts with the plank face down. The shear makes a small amount of tearout on the top side and a cleaner cut on the bottom. After making a cut, tiny fibers remain on the cut edge and are easily rubbed off with a bare hand. My uncle Wynn, a sometime carpenter, calls this "40-grit handpaper."

I have also used a pneumatic hand shear. But it seemed to need an awful lot of air, and my smaller compressor couldn't keep up with the demand. Snapper also makes a shear designed specifically for cutting curves, but I haven't been able to justify buying one—yet.

There are some limitations with the electric shear. It can't cut tight curves, and the shear can't cut all the way into the corner of a notch. But for special cuts, I use a carbide-grit blade on my jigsaw (see photo at left on p. 99). This process generates a small amount of dust, but nowhere near as much as a circular saw. Round holes, for light fixtures, can be cut with a carbide-tipped hole saw (see middle photo on p. 99).

Exterior electrical boxes are a little tricky to cut in if you're adding them after the fact because you must cut through both the cement siding and the plywood sheathing underneath. I cut the cement siding first using a small cordless circular saw with a carbide blade (see photo at right on p. 99). Corners can be cleaned out with a carbide rotary rasp in a cordless drill. Once the siding is removed for the box, I drill a hole and cut the plywood with a jigsaw.

# Finishing Up

Once all the siding is nailed up, all joints and ends have to be caulked with paintable latex caulk. When my painters do the caulking, they just run a finger over the butt joints to smooth the caulk. But if the siding has a grain pattern, I sometimes run a putty knife down the joint over the simulated grain so that the caulk follows the grain contours. When the siding is painted, the joint all but disappears (see photo below). Dents, chips, or cracks may be filled with any cementitious patching compound available at most hardware stores.

Fiber-cement siding is available unprimed and primed (see photo on facing page), and just recently, it became available painted. If siding is purchased unpainted, it should be primed and given at least one coat of 100% acrylic paint. Nails with colored heads are

**Caulk and paint make expansion gaps disappear.** Gaps at plank ends and butt joints must be left for expansion in fiber-cement siding. These gaps are filled with latex caulk, and a coat of paint makes the seams disappear.

available for attaching prepainted siding. Fiber-cement siding does not require back-priming, and cut ends do not have to be sealed if they are caulked properly.

In a couple of places on this project, I had to use narrow strips of siding under windows. To avoid having large roofing-nail heads showing on these strips, I ran a bead of construction adhesive along the backs of the strips and then tacked them in place with a pneumatic brad gun. The brads held the strips until the glue cured, and their tiny heads disappeared beneath the surface.

*John La Torre Jr.* is a carpenter with 25 years of experience. He lives in Tuolumne, CA.

## Sources

**FIBER-CEMENT SIDING COMPANIES**

**Cemplank**
(877) 236-7526
www.cemplank.com

**CertainTeed Corp.**
(800) 358-7164
www.certainteed.com

**James Hardie Company**
(888) 542-7343
www.jameshardie.com

**Maxi Tile**
(800) 338-8453
www.maxibuilding products.com

**OTHER SOURCES**

**Makita**
(800) 462-5482
www.makitausa.com

**Snapper**
Pacific International Tool & Shear Ltd.
(800) 297-7487
www.snappershear.com

**Primed and ready.** Made of portland cement, sand, and cellulose fiber, fiber-cement siding is resistant to all types of weather as well as insect damage, and it's fire-resistant to boot. The siding can be purchased primed, unprimed, or with a finish coat of paint.

# Installing Horizontal Wood Siding

■ BY FELIX MARTI

I've installed a lot of wood siding during my 20-plus years in the building trades. And whenever I'm given the option, I prefer to install the siding horizontally. On the aesthetic side, I like the way horizontal shadowlines emphasize the shapes of houses. But more important, horizontal siding lasts longer than vertical siding, especially in wet climates. That's because water running down a piece of vertical siding inevitably hangs along the bottom edge for a while, where it can be wicked into the end grain of the wood. The result: rotted wood, peeling paint, or both.

In this article I offer some suggestions for selecting wood siding, establishing a workable layout, and then installing the siding. Here in southwestern Colorado, we typically side a house with rough-sawn, unpainted red cedar, such as the channel-rustic siding shown in the photos. But the techniques I'll talk about are equally appropriate for painted clapboards on a colonial house.

The style of your house will strongly influence the siding pattern you choose and the grade of the material. For example, rough-sawn, knotty cedar or redwood evokes a rustic feeling. Clear, vertical-grain clapboards on a crisply painted colonial house, on the other hand, are more patrician. You can bet that patrician costs more than rustic when it comes to siding.

If you are budget-minded and if you plan to paint your siding, there is a relatively new composite product on the market called Inner Seal lap siding from Louisiana-Pacific. Inner Seal® is $\frac{7}{16}$ in. thick, and it comes preprimed in 8-in., 9½-in. and 12-in. widths. The pieces are 16 ft. long. I used Inner Seal siding on a house in the wet part of Oregon with good results. It paints beautifully, and it is stable and straight. I also like the fact that Inner Seal siding is top-nailed, which conceals the fasteners.

## Ordering Material and Checking for Defects

The pattern you choose influences the equation for determining how much siding you need. Measure the surface area of the walls to be covered, then subtract the square

The style of
your house will
strongly influence
the siding pattern
you choose and
the grade of the
material.

**The layout depends on the openings.** Full-width siding boards without notches will tuck nicely under the windows when the next two courses are applied. That's because the board at the bottom of the wall was ripped to a narrower width to simplify the cuts around the windows.

Unused finish is returned to its can by opening the end of the tank up and extending the plastic liner into a temporary spigot.

**DIP TANK AND DRYING RACK**
**Siding takes a bath in a long dip tank.** The author prefinishes cedar siding in a trough lined with cross-laminated polyethylene: 5 gal. of finish makes a bath about 2 in. deep. As one board soaks up the finish, another is held by a pair of brackets while excess finish drips off.

footage of the windows and doors. Refer to the chart on p. 107 to see what factor you need to multiply the footage by. Add 10% to this number for defects and waste, and you've got your siding total.

There are bound to be some defects in wood siding. Some are easy to spot, and some aren't. Some defects can be dealt with, and some can't. Discuss the grade of the lumber you want with your supplier, and agree before you place your order what is going to be acceptable.

Among the obvious defects are loose knots. Cut them out if they aren't too numerous, and use the resulting boards in places where shorter lengths are suitable. If loose knots are unacceptable in the grade you ordered, reject the material. In the premium grades, I think it's also fair to reject boards damaged by forklifts or the banding that holds the bundles together.

Warped or crooked boards often can be tamed with some additional fasteners. Bows should be cut out, and twisted or cupped

**Drying rack.** A pair of tilted 2x4s with 80d nails supports the siding as it dries.

boards should be rejected. Wane, the barky surface of the tree that sometimes shows up on the edges of boards, is always attached to sapwood. Depending on the grade you ordered, reject waney boards.

Checks are cracks in the ends or in the surface of the boards. Surface checking is caused by too-rapid drying and leads to stresses within the lumber: best to reject. End checking is common, but more than a few inches of check indicates too-rapid drying. Accept conditionally.

Fuzzy lumber probably was surfaced with too high a moisture content. If the rough surface is to be the exposed face, go ahead and use the board. If the smooth face is to be exposed, you can sand or plane the board smooth once it's dry. Boards with white specks or decay have been affected by fungus. Paint or stain will hide the specks, but you should cut out decayed portions. If the lumber won't be subjected to wet conditions, further decay won't occur.

Another problem that is virtually impossible to see is case hardening. This condition results from too-rapid drying and creates complex stresses within the lumber. Two signs of case hardening are binding of the sawblade and explosive splitting of the wood when you cut it. Ask for replacement pieces.

## Prefinishing with a Dip Tank

Both the California Redwood Association and the Western Red Cedar Lumber Association say you should prime, or seal, the back of your siding (called back priming) and finish it with a water-repellent mildewcide for a natural finish. If you put finish on just the exterior of a piece of siding, it can cup if the backside is exposed to moisture. So for a job like the one pictured here, I dip the boards in Penofin®, a linseed-oil-based clear finish. Penofin provides UV and water protection, and it offers a mildewcide treatment. To speed the process, I submerge my siding in a dip tank. For cedar siding to be painted, I prime it with a stain-blocking oil or latex primer. I prime other species of siding with acrylic latex primer.

My tank's bottom is a wooden I-joist (see photo at left on p. 104). The sides are ½-in. plywood screwed to the I-joist and to 1x2 stiffeners at the top. I made the ends out of 2x10 scraps. One end pivots for draining the tank (see photo at right on p. 104). I fastened a couple of drip brackets to the inside of the upper edge of the tank. After dipping a piece of material for 15 seconds to 20 seconds, I set it into these brackets so that the excess Penofin can drip back into the tank. While this piece drips, I fetch another, submerge it and leave it in the trough while I move the previous piece to the drying rack. At first I lined my tank with 6-mil poly, but it wasn't up to the attacks of ragged grain or the ever-present staples in the siding. So I switched to a cross-laminated polyethylene liner called Cross-Tuff. Cross-laminated polyethylene is amazingly tough and tear resistant. It's typically used under slabs and in walls as a vapor barrier.

My drying racks consist of a pair of cheapo sawhorse brackets, four legs and two predrilled 2x4s with gutter spikes in the holes (see photo on p. 105). Placing the dripped-dry piece of siding upside down in the rack will prevent beading of the finish on the bottom, exposed edge. This step is really only a concern if you're using a heavy-bodied stain.

As I dip the boards, I study each one for defects. When I encounter boards bad enough to be returned, I set them aside without dipping. After the boards have dried on the rack, I sort them into completely good boards, boards with some defects, and boards that must be cut into short lengths.

The best boards go near the entry and where the weather does its worst: the south side and the lower courses of the house. Okay boards go into less visible locations, where they are protected from the weather. The worst stuff goes way up high, out of sight, and out of the weather.

## Nailing Recommendations

Nailing patterns vary depending on the profile of the siding. But no matter what the pattern, one thing remains the same: Don't nail through overlapping pieces. To do so will eventually split the siding as it seasonally expands and contracts. Nails should penetrate at least 1½ in. into studs or blocking (1¼ in. for ring-shank or spiral-shank nails). Spacing should be no more than 24 in. o.c. Use box or siding nails for face nailing and casing nails for blind-nailing.

## Estimating Coverage

To calculate the amount of material required to side a house, first figure the square footage of the walls minus any openings. Add 10% for trim and waste. Now multiply your answer by the appropriate board-foot factor or linear-foot factor to tally the amount.

# Felt and Foam

I use 15-lb. felt building paper to protect the sheathing from moisture that might get by the siding. I don't use housewraps, such as Tyvek®, because I'm not convinced they're worth the extra money. What's more, 15-lb. felt makes a great background for the chalklines that I use to lay out the courses of the siding. I generally apply the felt as I go up the building. This technique saves me from ascending the wall twice, and I'm fairly certain I'll have the felt covered before the wind takes it off, or before rain or sun buckles it.

## BEVEL

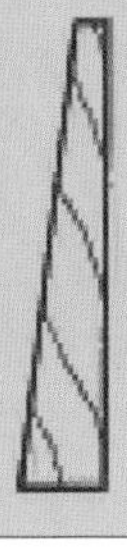

| | | | Nominal width | Dressed width | Exposed face | Factor for linear feet | Factor for board feet |
|---|---|---|---|---|---|---|---|
| **6 in. and narrower** | | **8 in. and wider** | | | | | |
| | 1 in. overlap. One nail per bearing, just above the 1 in. overlap. | 1 in. overlap. One nail per bearing, just above the 1 in. overlap. | 4 | 3½ | 2½ | 4.8 | 1.6 |
| | | | 6 | 5½ | 4½ | 2.67 | 1.33 |
| | | | 8 | 7¼ | 6¼ | 1.92 | 1.28 |
| | | | 10 | 9¼ | 8¼ | 1.45 | 1.21 |

## SHIPLAP (Dolly Varden)

| | | | Nominal width | Dressed width | Exposed face | Factor for linear feet | Factor for board feet |
|---|---|---|---|---|---|---|---|
| **6 in. and narrower** | | **8 in. and wider** | | | | | |
| | One nail per bearing, 1 in. up from bottom edge. | One nail per bearing, 1 in. up from bottom edge. | 4 | 3½ | 3 | 4 | 1.33 |
| | | | 6 | 5½ | 5 | 2.4 | 1.2 |
| | | | 8 | 7¼ | 6¾ | 1.78 | 1.19 |
| | | | 10 | 9¼ | 8¾ | 1.37 | 1.14 |
| | | | 12 | 11¼ | 10¾ | 1.12 | 1.12 |

## CHANNEL RUSTIC

| | | | Nominal width | Dressed width | Exposed face | Factor for linear feet | Factor for board feet |
|---|---|---|---|---|---|---|---|
| **6 in. and narrower** | | **8 in. and wider** | | | | | |
| | One nail per bearing, 1 in. up from bottom edge. | Use two siding or box nails, 3 in. to 4 in. per bearing. | 4 | 3⅜ | 3⅛ | 3.84 | 1.28 |
| | | | 6 | 5⅜ | 5⅛ | 2.34 | 1.17 |
| | | | 8 | 7⅛ | 6⅞ | 1.75 | 1.16 |
| | | | 10 | 9⅛ | 8⅞ | 1.35 | 1.13 |

## DROP

| | | | Nominal width | Dressed width | Exposed face | Factor for linear feet | Factor for board feet |
|---|---|---|---|---|---|---|---|
| **6 in. and narrower** | | **8 in. and wider** | | | | | |
| | 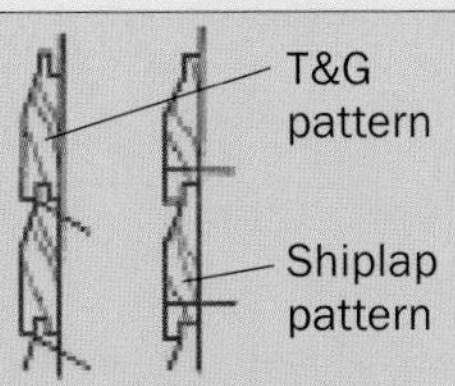 T&G pattern — Shiplap pattern. Blind nail T&G patterns; lace nail shiplap patterns, 1 in. up from bottom edge. | 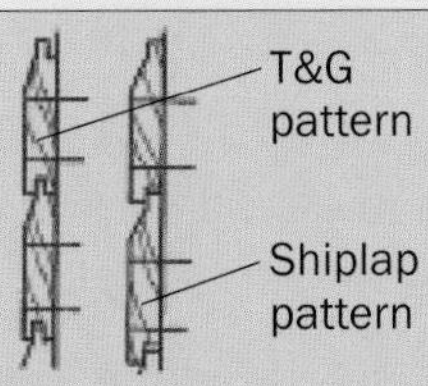 T&G pattern — Shiplap pattern. Two nails 3 in. to 4 in. apart to face nail, 1 in. up from bottom edge. | 4 | 3⅜ | 3⅛ | 3.84 | 1.28 |
| | | | 6 | 5⅜ | 5⅛ | 2.34 | 1.17 |
| | | | 8 | 7⅛ | 6⅞ | 1.75 | 1.16 |
| | | | 10 | 9⅛ | 8⅞ | 1.35 | 1.13 |

If you're going to apply wood siding over rigid-foam sheathing, remember that foam sheathing has no nail-holding ability. Use nails that are long enough to penetrate through the foam and 1½ in. into the studs.

## Two Poles Tell the Layout Story

I use a layout pole and a story pole to lay out the courses of siding before I cut a single board. The layout pole has saw kerfs along its length that mark the distance between each course of siding (see photo below). I find this distance by measuring a dozen boards to determine their average width.

The industry standard is to leave a ⅛-in. gap between siding boards. But here in southwestern Colorado, it's a rare board that doesn't shrink. That being the case, I leave a ¹⁄₁₆-in. gap between pieces of siding. Once the boards shrink, I'll end up with the recommended spacing. So the marks on my layout pole represent the average width of a piece of siding plus ¹⁄₁₆ in.

The story pole shows where window and door trims will be and where the siding starts and stops. To determine the best layout for the siding, I place the layout pole alongside the story pole, and I slide it up or down to see where the siding will break on window and door trims. The goal here is to avoid narrow rips of siding above or below a window or door. I rarely discover a sublime layout, but I usually identify the difficult areas before climbing a ladder with siding in hand. Sometimes I'll discover that the best way to avoid a nasty string of notched siding boards is to start the first course with a row of siding that has been ripped down to a narrower dimension (see photo on p. 103). Another strategy that can help avoid a string of complicated notches is to use a wide trim board above or below the windows.

Next I go around the house and jot down the rough measurements between the corners of the building and the openings, such as doors and windows. This list helps me determine how best to use the short lengths of siding that accumulate as the job progresses.

Once I've decided on the elevation of the siding on the wall, I use my layout pole to guide the marks for my chalklines. I use a water level to transfer the marks from one side of the wall to the opposite, and then around the corner and down the next wall.

**Layout pole.** The author uses a water level to locate the bottom courses of siding on all the walls. Then he marks the subsequent courses with the help of a 12-ft. stick with saw kerfs that correspond to the tops of the siding boards.

# You've Got Options at the Corners

At outside corners, I prefer to butt the siding to corner boards that stand a bit proud of the siding (see drawing at right). I typically use rips of siding to make the corner boards, and I fur them out with strips of plywood or rips from framing-lumber scraps.

Applied corners are similar to butted corners except that the installer can be fairly rough with end cuts. Because the corner is applied over these cuts, the ends of the boards will never be seen. This trim detail is much faster, but it has its drawbacks. Where the corner boards overlap certain kinds of siding, such as channel-rustic or beveled siding, you can create hundreds of little nooks for wasps, bees, spiders, and wind-borne grit. And it isn't easy to get paint or stain in there, either. If you choose this corner detail, be sure to seal the bottom of the corner so that it doesn't become a vertical gallery for mice or insects. Incidentally, you've got the same option at doors and windows as you do at corners. You can put casings around windows and doors (my preference) and then run the siding to them, or you can apply casings over the siding.

Mitered corners are another option, but my advice is to avoid them. They are supremely time-consuming, and even when done right, they will eventually open up as the weather does its seasonal work on the wood.

At inside corners I butt the siding to a corner board that is square in section. Applied inside-corner boards are faster, but they don't look as good and invite the same problems as applied outside-corner boards.

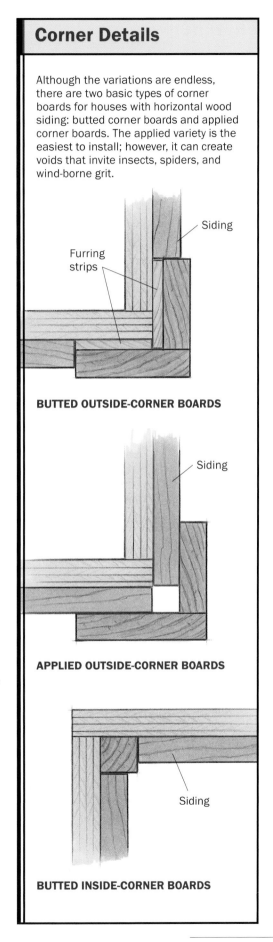

## Corner Details

Although the variations are endless, there are two basic types of corner boards for houses with horizontal wood siding: butted corner boards and applied corner boards. The applied variety is the easiest to install; however, it can create voids that invite insects, spiders, and wind-borne grit.

Siding

Furring strips

**BUTTED OUTSIDE-CORNER BOARDS**

Siding

**APPLIED OUTSIDE-CORNER BOARDS**

Siding

**BUTTED INSIDE-CORNER BOARDS**

## Fasteners

Within budget constraints, use the best fasteners available, especially in wet climates. My fasteners of choice are screws. I've used both square-drive stainless-steel and Phillips-drive brass flathead screws to affix siding to houses over ½-in. plywood sheathing. When the sheathing is this thick, I don't bother to break the siding over a stud because the screw threads get such a good bite into the plywood. I predrill all holes in the siding, which eliminates splitting, and I'm delighted in the resulting totally random siding pattern. I don't believe it takes much longer to screw siding in place than to nail it by hand. It probably costs no more if you consider the savings in material by not trimming the siding to break on a stud. The better lumber yield compensates for the extra time. What's more, using screws makes it a lot easier to replace damaged siding or to open a wall for remodeling.

Siding nails, in my order of preference, are stainless steel, hot-dipped galvanized, and aluminum. Spiral or ring-shank nails hold best, and you can generally count on dropping one size in nail length when you use them. The chart on p. 107 shows the recommended nail sizes and patterns for different types of siding.

In my experience, pneumatic nailers can too easily overdrive siding nails. The driver promotes splitting, especially at board ends, which gives water an easy access point. This point said, I must confess to using a ½-in. pneumatic crown stapler with 2-in. stainless-steel staples on my most recent rustic-siding jobs. The siding is ⅞ in. thick, and I ran staples into all the studs and the midpoints between them. The fasteners go in with the speed that makes pneumatic nailers so valuable, and they are virtually invisible on the fuzzy surfaces of rough-sawn cedar siding.

Before committing to using staples, I made several test installations and tried to remove the pieces of siding. The fasteners hung on for dear life, and the boards broke before the staples pulled through the siding or out of the studs. Be aware, however, that staples do not receive official sanction from the trade associations that represent redwood- and cedar-siding producers.

## A Three-Man Crew Is Just Right

Putting up siding is almost like painting siding: All the prep work seems to take forever, then the job rolls along at a good clip. Three people on a crew is optimum. One person can cut the siding to length while the other two take measurements, snap chalklines, and fasten the siding. A two-person crew is also efficient, and except for the huge amount of

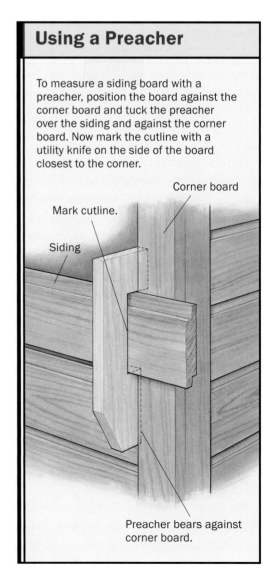

**Using a Preacher**

To measure a siding board with a preacher, position the board against the corner board and tuck the preacher over the siding and against the corner board. Now mark the cutline with a utility knife on the side of the board closest to the corner.

Corner board

Mark cutline.

Siding

Preacher bears against corner board.

climbing up and down, one person can get it done. A carpenter working solo can support the far end of a piece of siding with a J-shaped hanger made of baling wire or use a coat hanger.

I use a Hitachi® C8FB sliding saw for making most of my cuts on a siding job. I mount the saw in the middle of a long 2x12 that I place across a couple of sawhorses. Extension tables and fences flank the saw, supporting long pieces of siding during cuts (see top photo).

For acute-angle cuts that are beyond the swing of the Hitachi, I use a shooting board to guide my circular saw (see bottom photo). I screw my shooting board to a pair of fixed fences on a temporary bench top. The fences are ¹⁄₁₆ in. thicker than the siding, which allows me to slip each piece of siding under the fixture to make my rake cuts.

Start installing the siding with the bottom board, ripped to whatever width you determined by way of the layout stick and the story pole. The top edge of the board should align with your chalkline. I like the pieces of siding to abut the corner boards snugly, but I don't force them into place. Forcing the boards will push the corner boards out of alignment.

A piece of wood with a notch cut in it, sometimes called a preacher, is a superb little tool for marking exact lengths (see drawing on facing page). But using a preacher takes more time than simply measuring because you must lift the siding into place, mark it (use a utility knife), take it back down to the saw bench, and then put it back up. A remodel job that has out-of-plumb corner boards is the perfect place to use the preacher because the cuts won't be quite square.

An accurate, portable cutoff rig. A sliding-compound saw, such as the Hitachi C8FB, can make precise crosscuts in siding up to 12 in. wide. Long extension tables support the stock.

Shooting board for acute angles. A straightedge screwed to a pair of fences guides a circular saw through repetitive cuts that are beyond the swing of the sliding compound saw.

As I install the siding, I use a urethane caulk to seal the butt joints wherever the siding boards abut a window casing, a door casing or a corner board. I also use the urethane caulk on the back of the siding to secure any loose knots. The caulk will glue the knots in place.

**Keep water out at the joints.** The author uses narrow strips of waterproof membrane to prevent water from working its way behind the butt joints in the siding.

If I'm using nails or staples to attach the siding, I mark the stud layout on the tar paper with chalklines. Then I break the siding over a stud at butt joints. At each joint, I put a narrow strip of self-adhering waterproof membrane behind the joint (see top photo). The bottom end of this material is led out and over the top edge of the previous piece of siding, preventing water from going through the joint and behind the siding. Tar paper is often used for this detail, but I prefer waterproof membrane because it's less likely to become brittle and crack at the crease. I trim the little flap that protrudes below the siding with a utility knife. Inevitably, there are protrusions such as hose bibs, light fixtures, and outlet boxes that penetrate the siding. These situations are good opportunities for trim carpenters to show how fastidious their scribe cuts can be. We mark the cut with the siding held in place, and we use a jigsaw to notch the siding. A bead of urethane caulk seals the edges (see photos below).

**Weatherize the penetrations.** The holes made by hose bibs, pipes and junction boxes need to be sealed just like doors and windows. This outlet box has a metal flashing folded over its top, plus a flap of waterproof membrane to seal the joint at the wall. A bead of urethane caulk seals the sides and bottom.

I don't like to see a row of butt joints on a wall, so I make sure to stagger them by at least three rows on the same stud. Try to avoid butt joints near an entryway, and never butt trim boards over an entry. This advice would seem obvious, but I see the wrong technique over and over again. Do the splice somewhere else if the trim can't be done with one piece, and if a splice is needed, do it with a scarf joint.

## Trouble Spots Need Flashing

Unless they are protected by roof overhangs, the head casings atop windows and doors should be sealed against the weather with a flashing. I prefer to use copper flashings, but they're pricey. If copper flashings are beyond the budget, galvanized sheet metal, though bright and ugly, is a lot less expensive. These days, there is quite a selection of colors in baked-enamel sheet metal. If the job budget won't buy copper, it will usually accommodate the baked enamel, such as the ones we used on this job.

Where a roof meets a sidewall, you have several options. First, let the roofer roof, then side over his step flashings. Run the building paper over the flashings, but don't put any nails through them. If you do nail the flashings, you make it impossible for a roofer to jockey the flashings around at reroof time.

Your second option is to put up the siding before the roofing, making certain your fasteners are 2 in. to 3 in. up from the bottom of the siding and not so tight that the roofer can't slip his flashing under the siding. The third choice is to fasten a strip of counterflashing in place, side over it, and let the roofer slip his flashing under yours.

Siding often suffers at this intersection, usually because it has been applied too near the roof. Capillary action can draw water up

**Divert runoff away from sidewalls.** A gable roof that abuts a sidewall can direct water behind the siding at the junction along the roof's edge. Thwart the problem with a flap of flashing that directs runoff away from the wall.

**Sources**

**Louisiana-Pacific**
(800) 648-6893
www.lp.corp.com

**Manufactured Plastics**
(719) 487-7373
www.mpdplastics.com
*Cross-Tuff*

**Performance Coatings**
(707) 462-7333
*Penofin*

between the siding and the wall. This process can stain the siding, and it can promote rot. Leave a minimum of 2 in. between the siding and the roof. More if possible, say 4 in. to 5 in., especially if the roof is exposed to heavy weather. The greater clearance is insurance against saturation from mounded snow, soggy leaves, and the buildup from additional roofs.

If your sidewall continues beyond the roof's eave, fasten a piece of flashing along the roof's edge to keep water from going behind the siding (see photo above). Confer with your roofer before proceeding with any of these options because he's the one who will have to stand behind the roof. And make sure that both you and the roofer are using the same kinds of flashing materials. For example, don't mix copper and galvanized.

*Felix Marti designs and builds home with an emphasis on energy efficiency and low maintenance. He lives in Ridgway, CO.*

# Rain-Screen Walls: A Better Way to Install Siding

■ BY MARK AVERILL SNYDER

sensed trouble as soon as I saw the house. I was there to bid on painting the eight-year-old, multimillion-dollar home. What first struck me was the sheets of paint dangling from the cedar siding (see photo at left on facing page). Even more alarming, the paint peeling from the walls was from the house's third paint job.

When I learned that the previous paint job was done only 2 years before, I knew that the problem ran deeper than the paint. Jittery about hiring yet another painter, the homeowners agreed when I suggested calling in a consultant, Joseph Lstiburek.

When Joe saw the house about a week later, he said, "I'll tell you in two minutes what's wrong if we can pull one clapboard." Pulling off one clapboard revealed tea-colored Tyvek housewrap, soaked like blotter paper (see inset photo on facing page). Both the clapboards and the plywood sheathing were saturated. I was stunned.

Joe and I checked for possible moisture sources, from the attic to the basement. There was no evidence of any interior moisture source. Furthermore, we found no flashing deficiencies that could lead rainwater behind the housewrap. The pattern of wetting, coupled with the lack of any interior moisture source, pointed to rain that found its way past the siding and the housewrap.

None of this surprised Joe, who explained the science behind the failure and the reason why the one sure way to correct it was to build a rain-screen wall.

## Isolate the Housewrap from the Siding

Rain-screen walls have an airspace between the siding and the building paper or housewrap. (Installed behind a rain screen, plastic housewraps can work. I prefer 15-lb. tar paper because of its century-old track record. Either one makes up what is called the drainage plane). This airspace is created by applying vertical furring strips over the

drainage-plane material. The siding, typically clapboards, is nailed to these strips.

The airspace does several duties. First, it spaces the siding away from the drainage plane, minimizing any chance of extractive bleed (see drawing on p. 117). Back-priming the siding would do the same thing, and it is a good practice. But it does nothing to address capillary action. As demonstrated by the grade-school experiment of putting a stalk of celery into a glass of colored water, capillary action can draw water upward along a continuous surface. Water can move long distances in this way, so a raindrop climbing the back of one clapboard to the top of another, then to the wall sheathing, is not an imaginary event.

The airspace in a rain-screen wall provides a capillary break, stopping water driven behind the siding by capillary action from ever reaching the drainage plane. Vented top and bottom, the siding has a way to dry from the water that gets behind it. Finally, this vented airspace helps equalize the air pressure on each side of the siding, reducing the chance of wind-driven rain being forced or drawn behind the siding.

You might argue that rain should never get past a properly detailed, standard siding job. I once would have argued the same. However, when my crew stripped this house's walls of their trim and clapboards,

## Sources

**Samuel Cabot, Inc.**
(800) 877-8246
www.cabotstain.com
*Cabot's Problem Solver*

**Cor-A-Vent**
(800) 837-8368
www.cor-a-vent.com

**GAF**
(800) 234-4285
www.gaf.com

**Grace Construction Products**
(800) 444-6459
www.graceconstruction.com

**Benjamin Obdyke, Inc.**
(800) 346-7655
www.obdyke.com
*Cedar Breather*

**Pratt and Lambert**
(800) 289-7728
www.prattandlambert.com

Tannins leaching from the back of the cedar siding, which had not been back-primed, ruined the house-wrap's water repellency (facing page). Any rain that got past the siding soaked the sheathing, creating a reservoir that kept the siding wet and the paint peeling (left).

we stood back and marveled (see inset photo on p. 114). The worst damage was around door and window openings and under clapboard joints, exactly where water had leaked under the unprimed siding. The funny thing is that even though we looked carefully at the siding's installation, it was hard to see how water had leaked in. The workmanship was perfectly acceptable.

## Layers of Membranes, Flashing, and Tar Paper Protect the Sheathing

The existing clapboards, trim, and housewrap of this house had to be removed, and damaged sheathing and framing repaired. Once these tasks were done, the only difference between this job and a new house was that I was locked into existing window and door trim, and soffit details.

Had this been new construction, I would have furred out the door and window units, thereby keeping all the trim on one plane. The 84 windows and 18 doors were already installed with their 5/4x4 casings applied against the original housewrap. The homeowners drew the line here; they did not want these units removed. The best that my crew could do was to pry the casings away from the house enough to slip double-thick 15-lb. tar-paper splines under the casings.

"Be the raindrop," an old-timer once told me. Although this phrase might sound like a hazy bit of Eastern philosophy, imagining the path of a raindrop is key to understanding waterproofing details. The idea is to start at the bottom and layer upper flashings and drainage planes over those below.

On a new home, the first layer would be a pan of self-adhering bituminous membrane at the rough sills of doors and windows. I use Grace Ice & Water Shield. There were only a few doors on this house that had to be taken out to repair the surrounding framing, so we could use this detail only in these places (see photo at left).

We also adhered a 3-ft.-wide course of Ice & Water Shield to the sheathing at the bottom of the walls, all the way around the house, to guard against dripping water splashing back to the house.

Other spots that we covered with Ice & Water Shield were walls that rose above an intersecting roof. On a new house, we would have extended half the width of the Ice & Water Shield onto the roof deck and half up the wall. This option was closed to us on this house because we weren't tearing off the roof. The best that we could do was to bend down the existing step flashing to expose the wall sheathing, which we then covered with Ice & Water Shield.

The next layer in the rain screen is the cap flashings above windows, doors, and other wall penetrations such as dryer vents and light-mounting blocks (see photos on p. 118). These flashings are installed with

**First, flash vulnerable spots such as rough sills to drain water.** Note how the tar paper to the right of the opening laps over the top of the Ice & Water Shield flashing. In turn, the Ice & Water Shield laps the tar paper below.

# Detailing a Rain-Screen Wall

Furring strips space the clapboards away from the wall, creating a vent space that helps dry the back of the siding. Layered tar paper and flashings behind the furring strips create what is called a drainage plane, which alone would protect the house from water, even if there were no siding. This belt-and-suspenders approach can contribute to the longevity of both the house and its paint.

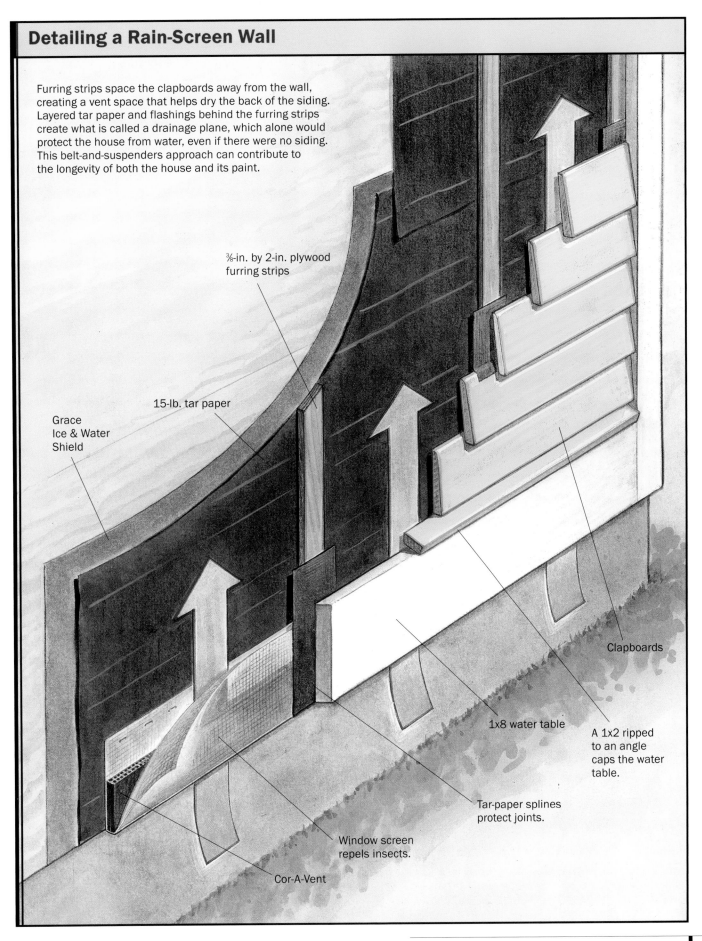

⅜-in. by 2-in. plywood furring strips

15-lb. tar paper

Grace Ice & Water Shield

Clapboards

1x8 water table

A 1x2 ripped to an angle caps the water table.

Tar-paper splines protect joints.

Window screen repels insects.

Cor-A-Vent

**Cap flashing protects door and window casings.** Note how tar paper covers the furring strip by the side casing, and how the cap flashing and tar paper are all layered to drain down.

**siding leaks at joints, let it drain.** The tar-paper splines under this light-ounting block both cover the furring strips and extend over the next lower pboard to drain water away from the house.

vertical flanges over the tar-paper splines that we slid behind the casings directly against the wall sheathing.

I used copper and lead flashings. Where the flashing would show, I used copper. Where it wasn't particularly visible or where I needed a flexible flashing, I used lead.

## Tar Paper Has Worked for a Century; Why Change?

The final step before nailing up the furring strips was to wrap the house in 15-lb. tar paper. Recently, I've switched to Shingle Mate® by GAF, a fiberglass-reinforced tar paper that stays flat and doesn't pucker up when it becomes wet as regular tar paper does. We started at the bottom, covering the layer of Ice & Water Shield. We applied the tar paper horizontally, lapping each succeeding course a minimum of 4 in. over the lower course. This layer of tar paper also laps both the step flashings at roof intersections and the cap flashings over doors, windows and the like (see top photo).

We made the furring strips from ⅜-in. AC fir plywood. We ripped most of the strips 2 in. wide, plus a few 6½ in. wide for under corner boards. (The corner boards were to be ¾ in. by 5 in., so that left 1½ in. of furring strip to support the abutting siding.) On new construction, we would have furred out the windows and doors, but as I said before, the owners didn't want us pulling and rein- stalling all those units.

As luck had it, the door and window casings were of 5/4 stock. This good fortune allowed us to add the ⅜-in. spacer parallel to the side casing and still have the clapboards be flush with the casings. For added protec- tion, we covered the furring strips next to the window casings, a spot we knew from the tearoff to be vulnerable, with tar paper.

To provide the greatest possible chance for water behind the siding to dry out, the tops and bottoms of the channels between the furring strips were left open to the air. We screened these channels' bottoms and tops to prevent bugs from entering. This step was a simple matter of stapling 12-in. wide strips of screen along the bottom edge of the building so that 6 in. of it hung below the bottom of the sheathing (see drawing on p. 117). We then nailed the furring strips, with their bottoms flush to the bottom of the sheathing, over the screen. The furring strips were nailed to the house directly over studs with stainless-steel 7d ring-shank nails.

I wanted to back up the water table so that it didn't flex if someone leaned on it, yet still allow ventilation. To accomplish this, we filled the spaces between the furring strips' bottoms with Cor-A-Vent, a plastic roof vent that resembles corrugated cardboard. We ripped the Cor-A-Vent to 3 in. wide and peeled its layers apart so that they matched the furring strips' thickness.

We repeated this detail at the tops of the walls, finally folding the screen over the furring strips and Cor-A-Vent, top and bottom. The Cor-A-Vent fills the void between furring strips; it's rigid and waterproof, and it supports the screen so that bugs can't crawl in.

Above the windows and the doors, we ran a horizontal furring strip over the flange of the cap flashing. To vent these spaces, the vertical furring strips stop ½ in. shy of the horizontal strip.

The final step on this job before beginning to nail on trim and clapboards was to install counterflashing where roofs met sidewalls. We used 8-in.-wide lead, nailed to the furring strips and bent 90°, to go 3 in. over the cedar roof shingles. This flashing was crimped down over the butts of the shingles to minimize the intrusion of wind-driven rain (see photo at right).

# Even on a Rain Screen, Back-Priming Is Good Practice

After we built the rain screen, skimping on the trim and the clapboards made no sense. I have found that of the commonly available woods, only two are suitable for harsh climates such as New England's: cedar and redwood. With a minimum amount of maintenance, these two woods will last a lifetime. We chose cedar.

I sanded and dusted all six sides of every trim board that I used. This task may seem to be overkill—it's certainly tedious—but I accomplish several ends at once. The process allows me to read the boards for cups and cracks, dings, and dents; it removes dirt and wax film from the board; it levels planer marks; and it scuffs up the surface for superior paint adhesion.

> To provide the greatest possible chance for water behind the siding to dry out, the tops and bottoms of the channels between the furring strips were left open to the air.

**Counterflashing covers step flashing, Ice & Water Shield and tar paper. The counterflashing is layered over the lower clapboard to drain, while the upper clapboard stops above it to defeat capillary action.**

After sanding, I primed all surfaces of every board with an alkyd primer, Pratt and Lambert® Suprime® 8. Alkyd primer takes at least 48 hours to dry on cedar, so I always needed to be 2 days to 3 days ahead of the installation crew.

Because I could buy clear, preprimed cedar clapboards, I avoided the tedium of sanding them. To be sure, though, we primed every cut end before installation.

## Siding Goes on Conventionally

The first trim boards up were corner boards. Before nailing up the corners, we covered their furring strips with 15-lb. tar paper.

Next came the water table, a 1x8 capped with a 1x2 that we ripped to pitch 15° to drain water outward (see drawing on p. 117).

The water table is installed level with the bottom of the plywood sheathing. We avoided joints in the water table as far as was possible. Where they were unavoidable, we scarf-joined the boards and protected the furring strip behind the joint with a tar-paper spline.

The clapboards were installed conventionally, but on top of the furring strips. They were face-nailed with small-head 6d stainless-steel ring-shank nails. Every joint was backed up by a 3-in. by 5-in. tar-paper spline that laps the clapboard below. All the joints were tight, a razor-knife-blade thickness or less, but not so tight that they would bow the corner boards or window casings.

We lapped the lower tar-paper splines at interruptions such as windows or light-mounting blocks over the top of the clapboards. The final trim piece is a 1x4 installed in place of a final clapboard, tight to the sof-

## Venting the Top of a Rain-Screen Wall Unobtrusively

The air from below needs to exit somewhere. Rabbeting the trim board that caps the siding completes the path.

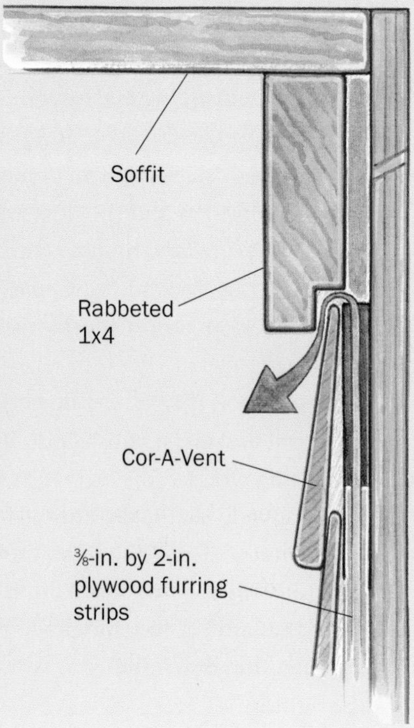

Soffit

Rabbeted 1x4

Cor-A-Vent

⅜-in. by 2-in. plywood furring strips

fit. This board's lower edge is rabbeted ¼ in., creating a weatherproof, almost-invisible vent (see sidebar on facing page).

Even on a rain-screen wall, good paint is worth the cost. I ordered clapboards pre-primed with Cabot's® Problem Solver. I spot-primed cut ends with Pratt and Lambert Suprime 8; then the entire house got three coats of Pratt and Lambert's best acrylic exterior paint, Accolade®, in an eggshell finish.

We'd made the joints tight enough that no caulk was needed. Caulk doesn't last, and it doesn't make up for poor carpentry.

I've been back recently to look over this house. After 3 years, the paint still looks new—on a house that previously wouldn't hold paint for half that time.

*Master carpenter and writer Mark Averill Snyder has been building homes for 25 years. He lives in Greensboro Bend, VT.*

# How Housewraps and Building Papers Fail

■ BY JOSEPH LSTIBUREK

The primary function of a housewrap (Tyvek and Typar® are two common examples) or building paper (15- or 30-lb. tar paper) is to drain rain that penetrates the siding through leaky joints or capillary action. Marketing claims to the contrary notwithstanding, housewraps do very little to reduce any air infiltration.

The problem with housewraps, and building papers to a lesser extent, is loss of water repellency. Contaminants, referred to as surfactants, degrade the water repellency of building papers and housewrap, allowing the wetting of the housewrap or building-paper surface by water. (To learn how surfactants work, visit www.chemistry.co.n2/surfactants.htm.) Water-soluble extractives in wood, such as tannins and wood sugars in Douglas fir, redwood, and cedar, as well as detergents and soaps, are surfactants. Once the surface is wet, pores in the housewrap or building paper fill, allowing liquid-phase water to pass through. Once water penetrates the housewrap into a wall, peeling paint and rotten siding often follow.

Back-priming wood clapboards and trim helps isolate surfactants in the wood from the housewrap or building-paper surface. Similarly, providing an airspace between wood trim and siding and the housewrap or building paper using furring strips (Cedar Breather® is a commercial product for this application) reduces the potential of surfactant movement. I recommend both practices.

Stucco should never be installed in direct contact with any plastic-based housewrap. Stucco can adhere or bond to the housewrap surface, allowing housewrap pores to become wetted and subsequently establish capillary flow. Also, many stuccos add surfactants to improve workability and freeze–thaw resistance. A drainage space between stucco and housewrap is essential to control water.

Stucco does not typically bond with building papers. However, most stucco applications over building paper result in insufficient drainage. I recommend using at least two layers of building paper under stucco to allow some drainage between the two materials.

Plain dirt can also cause housewrap to leak, just as Scotchgard™-coated fabric that becomes dirty must be cleaned and re-treated to re-establish water repellency. Don't let your housewrap become muddy or dirty.

*Joseph Lstiburek is a principal in Building Science Corporation (www.buildingscience.com) in Westford, MA, and the author of Builder's Guide to Cold Climates (published by The Taunton Press, 2000).*

# Siding with Cedar Shingles

■ BY OLIVER GOVERS

In one of my first siding jobs, a retired and very persuasive reverend had convinced me that his house could be sided with cedar shingles for only a little bit more than his neighbor's, a house that my helper and I had just completed. Newly self-employed and eager for work, it hadn't occurred to me that the reverend's house was half-again as large as his neighbor's. And all of the little extras that he wanted—at little or no extra charge—kept adding up. Only 2 days into the job, we were already 35% over budget.

That was several years ago, and I lost a lot of money and a bit of confidence on that job. But the reverend was pleased, and the house came out well. I'm better with estimating now; and despite the rough start, I'm still siding houses with shingles. Cedar is wonderful material to work with, has a lot of curb appeal, and is a practical siding choice.

When I side a house with shingles, I like to wrap the corners. I know that many East Coast houses with cedar-shingle siding are traditionally trimmed out with corner boards, but I'm not particularly fond of them. To me they seem like a shortcut, breaking up into many separate sections what should be continuous siding wrapping around the house. So I generally wrap my siding with woven corners (see photo on facing page), a practice that seems more common here in California; I think this style leaves a cleaner look and a more craftsmanly finish. But whether you weave your corners or butt your shingles into corner boards, cedar-shingle siding is a rewarding and low-tech approach to a traditional, high-quality exterior finish.

One thing that I like about siding with cedar is that it doesn't require an armory of specialized tools. A hammer, a utility knife, a tape measure, a chalkline, a block plane, a drill, and some levels are all I need. I'll occasionally use a table saw or a jigsaw, but I don't consider them to be necessities. A compressor and a pneumatic stapler can speed up work but also are optional.

Of course, unless you are shingling up only 5 ft. or 6 ft. on the wall, you'll also need scaffolding. On an extensive job I usually use the pipe scaffolding already set up on the site, or I'll rent from a supplier. Ladders and ladder jacks also work well in certain situations, as do pump jacks. Whatever

**Siding with shingles requires just a few basic hand tools.** Although a pneumatic stapler can speed things up, most shingling can be done with only a hammer, a utility knife, a few levels, a chalkline, and a block plane.

*Cedar is wonderful material to work with, has a lot of curb appeal, and is a practical siding choice.*

type of scaffolding you choose, though, make sure that it is safe. Siding work can take you to high places, and serious injury or death is a distinct possibility if unstable or makeshift scaffolding fails.

## Buy Good Shingles

Lousy shingles in the bundle are still going to look lousy on the house, so it's important to pay attention when selecting shingles at the lumberyard. I look for shingles with uniformly thick—about ⅜ in.—butts. Then I check the exposed faces of the shingles on both sides of the bundles to see that the grain is reasonably tight and as straight as possible, with no knots. These characteristics make hand-cutting and planing of the shingles more predictable and accurate. I also like to check the shingles for square. A good bundle of shingles will have a variety of widths: plenty of bed sheets (12-in.-wide

shingles), 6-in., and 9-in. shingles, and not too many narrow ones.

Red-cedar shingles are generally used here on the West Coast, and they typically weather to a fairly dark reddish brown color. White cedar is less common here; it generally weathers to a silver-gray. In my region the best-quality shingles are called #1 grade, and they are the ones I generally use because they are free of knots and have a much smoother surface than lower grades. I'm not shy about sending bundles back to the lumberyard that aren't up to snuff.

As I would do with any type of siding, I install a weather-resistant building paper over the sheathing. In the San Francisco Bay area where I work, which has mild to average year-round weather, 15-lb. asphalt-saturated felt has always worked well.

I also verify that all windows and doors are properly flashed. When installing the building paper, I pay particular attention to inside and outside corners. I find that creasing the felt before installing it makes for tight corners that won't be damaged when they are shingled.

## Choose the Right Fasteners

I like to use either stainless-steel or brass ring-shank nails for all shingles that will have exposed fasteners, such as the last courses of shingles under windows or eave, or on corners. Otherwise, in unexposed locations I use standard galvanized nails or staples. Fasteners should be long enough to penetrate at least ¾ in. into the wall. If I'm shingling over sheathing, 4d nails or 1½-in. staples work well. If I am shingling over gypsum board (required by some local building codes for fire protection), I then have to use longer fasteners.

When I'm doing different applications, I prefer using different fastening methods. When I want to accomplish volume production on wide-open sections, I often use a pneumatic stapler. When I'm shingling

**Choose corner shingles carefully.** Although both shingles came from the same bundle, the tight, straight grain of the shingle on the left will make it much much easier to trim with a utility knife or a block plane.

**Felt paper makes a good underlayment.** Economical 15-lb. asphalt-saturated felt comes in easily managed sizes, shows chalklines well, and helps keep moisture out of the sheathing.

small areas, working on inside and outside corners, and installing specially cut shingles, I prefer to nail by hand. In any case, two fasteners should be used per shingle, placed about ¾ in. from either edge and about 1 in. above where the butt ends of the following course will be.

## Careful Layout Keeps the Shingles on Course

On a typical two-story house, I will usually mark three level reference lines all the way around the house: one in the middle, one about 5 ft. up from grade and one about 5 ft. down from the soffit. Before I mark these lines, though, I need to determine exactly where I want the courses to be, which depends in part on shingle exposure. For example, when shingling a house using a 5-in. exposure, I might want the last course of shingles underneath the soffit to be a full 5 in. To do this layout, I need to measure down from the soffit and mark the reference line on a multiple of five. If I want the first course all of the way around the house (providing that it's on a relatively level grade) to be 5 in., I need to measure up from where the first course will be and mark all of my reference lines on multiples of five.

Once I've determined my layout and made reference marks on one corner, I generally use my trusted 6-ft. level to transfer these marks around the building. Another quick way to transfer these level marks from corner to corner is with a builder's transit or a water level. If you are confident that the building has level soffits, you could also use a story pole that has all of the course reference lines marked on it. Because these level lines dictate the layout of the job and help ensure that the courses are going to end in the same place where they started, I pay particular attention to marking them and chalking them accurately.

I usually base my layout decisions on the house's most prominent facade, typically the front. I take into consideration such factors as the visibility of the first and last courses, the relative height of doors and windows, and complicating factors such as porches or decks. Ideally, I like the bottoms of courses to line up evenly with the windowsills, the head casings, and the frieze, but this situation doesn't always happen. And when tying into another building, the shingle layout is predetermined by existing courses.

> **TIP**
>
> When installing the building paper, pay particular attention to inside and outside corners. Creasing the felt before installing it makes for tight corners that won't be damaged when they are shingled.

Of course, shingle siding can be installed using different exposures. Although the typical exposure is 5 in., I've seen houses with 7-in. exposures. I don't recommend that an exposure exceeds 6 in. because it is important to get at least double-layer coverage. A typical shingle is roughly 16 in. long, so using a 6-in. exposure will provide double-layer coverage, using a 5-in. exposure will provide triple-layer coverage, and so forth. With a 5-in. exposure I can usually plan on about four bundles of shingles per square (or 100 sq. ft.) of coverage.

## Installing the Starter Course

A typical overhang below the sheathing for the first course of shingles is 1½ in., but this overhang can be adjusted slightly to keep the starter-course exposure close to the exposure of shingles in the field. In some cases, however, the layout will leave no option except to have a short starter course (see photo below).

After the corners are built, fill in the field. A level can be used to mark the layout when the distance between corners is short; otherwise, a chalkline works well.

The starter course is typically built up from two layers of shingles, one on top of the other. The butt ends of both layers are flush with each other, and the sides (or vertical edges) of the second layer of shingles are offset from the sides of the first by at least 1½ in. The double layer of shingles on the first course prevents water from seeping underneath and into the sheathing. The double layer also builds up the proper thickness at the base so that the second course of shingles lies properly.

Although the starter course usually consists of a double layer of shingles, there can be exceptions. Sometimes three or more layers are used to flare the starter course. In other cases, more than two layers of shingles may be installed in a starter course to compensate for an out-of-plumb wall. Sometimes I have to experiment a little to find the right built-up thickness to produce a suitable starter course.

Because the starter course hangs below the sheathing, I can't snap a chalkline to mark its location. Instead, I stretch a stringline to keep the row of shingles on course. This stringline runs taut between nails driven into the butt ends of the first starter shingles on opposite corners of a section, or it is run between the corner boards and at the right elevation for the layout. I take care to install the shingles on this starter course so that they are just above this string and not resting on the string, which will keep them from deflecting the line and cause the course of shingles to sag.

## Weave the Corners First, Then Fill in Between

I've met a few shinglers who prefer to build woven corners a course at a time as they shingle a section. Although there is nothing wrong with doing it this way, I think that building all of the corners of a section at once and then shingling everything in

A woven corner makes a weatherproof joint. While the upper part of the corner shingle is trimmed past flush with a utility knife, the exposed edge is trimmed just barely proud with a block plane.

between is much more efficient. This way I already know where all of the courses will be, and all I have to do is chalk a line from one corner to another. Essentially, the layout work is done before I start shingling the field. I've always done inside and outside corners this way, but occasionally I'll do them a course at a time if both ends of a section are within arm's length or so.

Weaving a corner simply means installing the corner shingles so that the exposed edges face the same way on every other course. Alternating the butted and exposed edges of shingles on each course creates a tighter and more weatherproof joint. I choose potential corner shingles carefully, making sure that they are uniform in thickness and have a relatively straight grain, which will make cutting and planing them easier.

The first corner shingles on either end of a section are also part of the starter course and are installed in a double layer (see bottom photo). The next course begins on the predetermined exposure pattern, and this layout is then repeated right on up the corner. Remember, where the corner shingles are placed determines the exposure and the location of the courses in the field, so they must be installed with accuracy and with an eye toward the resolution of any anticipated layout problems.

I nail the first shingle to the wall with about 1 in. of the shingle extending past the corner. I use a torpedo level to check for level and fasten the shingle with three or four nails. Although the other shingles on the house will be installed using two fasteners per shingle, the extra fasteners on the top portion of the corner shingles will keep them from pivoting when they are either cut or planed.

*Weaving a corner simply means installing the corner shingles so that the exposed edges face the same way on every other course.*

Start at the corners. The starter course for woven shingles begins at the corners. The shingles are usually doubled, have vertical joints staggered by at least 1½ in. to prevent water penetration and hang below the sheathing another 1½ in.

Once the first shingle is installed, I use a sharp utility knife to trim back the excess, still leaving between ¼ in. to ½ in. of shingle to be trimmed. Starting at about 1 in. above the exposure line, however, I trim the shingle back past the corner slightly so that it's not flush with the corner. If I don't trim the shingle, the corners will start to bulge outward slightly, and this problem will compound itself with every successive course of shingles. I then use my block plane to plane the exposed edge of the shingle so that it's flush with the corner, being careful not to bevel the edge or to plane past flush.

Then I face-nail the shingle about 1½ in. up from the butt end and about the same distance in from the edge with a ring-shank, stainless-steel or brass nail. Face-nailing will keep the exposed part of the shingle tightly fastened to the wall so that it doesn't warp or curl later on.

After the first shingle has been leveled, fastened, cut, planed, and face-nailed, I can then install the second shingle, or the other half of the corner. I place the shingle against the wall just like the first, check it for level, make sure there is enough overlap at the corner, and then nail it home. I again use my utility knife to trim the shingle roughly to size, then finish it with my block plane. It's better to be one stroke of the plane shy of flush than it is to be one stroke past flush.

After planing the edge flush, I drill a pilot hole about 2 in. up from the butt end of the shingle and only about ¼ in. in from the edge and into the first shingle. I'm careful to drill the hole at a square or slight inward angle so that the nail won't emerge from the face of the first shingle. Nailing the shingles together in this way creates a waterproof corner that will remain tight over the years. Then I continue on up the corner, installing the shingles at the proper exposure and alternating the exposed edges so that the corner is woven.

Woven inside corners are built essentially the same way as outside corners, with a few minor variations. No face-nailing is necessary because the second shingle installed on an inside corner presses the first shingle tight into the wall. When cutting and planing the shingles on inside corners, I trim more material off the bottom than the top edge, which is just the opposite of outside corners. On both types of corners, I try to achieve a tight joint while I keep the butt ends of the shingles level.

# Shingling the Field

With prebuilt corners or with corner boards, filling in the field is pretty straightforward. A chalkline snapped from corner to corner at each course ensures that the courses remain nice and straight. I like to place the butt ends of the shingles so that they are just covering the chalkline, which helps conceal the chalk. Another option is to tack a straight 1x guide strip to the wall to mark the course.

Whether you run the courses from left to right or vice versa doesn't really matter. Also, I've found that leaving a gap between shingles, as is sometimes recommended, is unnecessary. The shingles I get from my local suppliers are always fresh and still contain a fair amount of moisture. The resulting shrinkage of the shingles after they're installed creates a slight gap, usually about $\frac{1}{16}$ in. to ⅛ in., which is plenty. If the shingles you're working with are bone dry, you may want to space them $\frac{1}{16}$ in. to ⅛ in. apart, which would allow them to expand with moisture. The only time that I install shingles with a gap is when I'm trying to match the gap of any existing shingling.

Sometimes shingles need to step up. Where there is a set of steps or a foundation on a slope, shingles should be stepped up and over. Remember to maintain a double layer on the bottom course and a consistent vertical and horizontal overlap of the foundation.

Choosing which shingle follows which when shingling the field is largely a matter of taste, as long as you remain within a few broad guidelines. First, you should avoid aligning joints vertically and try to maintain at least a 1½-in. overlap between joints. Second, try to vary shingle widths so that the shingles on the wall look random. And as my old boss used to say, "Never put a fat next to a skinny." At first I thought he was referring to the width of the shingles, and in a way he was. But what he was really talking about was the thickness at the butt of the shingle because this dimension can vary considerably from the nominal ⅜ in.

I'll occasionally have to trim a bad edge on a field shingle with my utility knife or smooth it with my block plane. Usually the last shingle in a course needs to be trimmed to size, too, which is also easily done with the knife and the plane.

## Starting a Course on an Uneven Grade

Not all houses are built on level lots, so sometimes a starter course will have to follow a significant grade. Starter courses on a grade can be installed one of two ways: by stepping the courses up the grade or by running the shingles "into the ground" and then cutting them parallel to the grade.

If the foundation of a house adjusts to the grade by having a series of steps running up the grade, I'll simply step the courses up the grade, overlapping the foundation horizontally and vertically by about 1½ in. (see photo above). If the foundation of a house runs up the grade continuously without stepping, then it will be necessary to cut the courses of shingles parallel to the grade.

After the courses have run up the grade and I'm able to run courses from end to end of a section, I'm ready to trim off the excess shingles. I'll snap a chalkline roughly parallel with the grade, being careful that the shingles overlap the foundation by about 1½ in. With a circular saw set just deep enough, I'll then trim the courses to the chalkline.

## Adjusting the Exposure

As I'm shingling the field between the corners, sometimes I'll need to adjust the shingles' course exposure. Out-of-level windows or doors may cause problems. Do you keep the courses of shingles level or run them parallel to the out-of-level windows or doors? I've found that it is usually best to split the difference and run a slightly out-of-level course to minimize the height discrepancy of the course of shingles.

Whether the courses go slightly off intentionally, as I just described, or whether I discover that they are off by double-checking the courses to reference lines, they need to be made up. Depending on the length of

the course, irregularities of more than ¼ in. should be made up in small increments. For example, if one end is ¾ in. low, make it up over the next six courses, if possible, adding ⅛ in. to every course on the side that is low, rather than all at once. The idea is to make adjustments gradually so that the eye cannot pick them up.

Shingling around windows and doors is pretty straightforward with only a few things to keep in mind. The shingles on the last course under any windows or doors should be cut with a saw, rather than a utility knife, to keep a true edge and maintain a clean look. If the window or door is out of level at all, adjust the courses slightly to stay parallel with the head casing.

The shingles at the bottom and top corners of windows and doors need to be cut around these corners so that there is no joint in vertical alignment with the edge of the trim. This process is referred to as notching, or "L"-ing, which describes the shape of the shingle after it has been cut to fit the corners (see photo below). And filler shingles need to be installed above the casing to build out this area to the proper thickness. There will usually be some shingle tops left over from cuts I've already made on the table saw, and these scraps are useful for this purpose because it makes use of wood that might otherwise be thrown away. Once again, the idea is that the proper built-up thickness is maintained wherever a course is started or, in this case, stopped underneath or above a window or door.

## Dormer Shingles Can Be Precut

Shingling a dormer or a wall over a roofed bay window is similar to running shingles up a grade with one exception: The shingles need to be precut to match the pitch of the roof before they're installed, rather than cut in place afterward. First I make sure that the cut angle accurately matches the pitch of the roof. After verifying this, I set my miter

**Notch the shingles around windows and doors. Because water tends to penetrate vertical seams, shingles around casings should be cut in an "L" shape to overlap this vulnerable joint.**

Shingling up a slope. When shingling a dormer or around a bay-window roof, the author precuts a number of shingles to the same pitch as the roof slope, but in a variety of lengths. These shingles can then be used to maintain the proper built-up thickness for succeeding courses.

gauge on my table saw and precut a series of shingles at three different points so that I'm left with three different shingle sizes: short, medium, and tall.

When I run a course from the corner and back toward the roof and begin to need angled shingles, I use the tallest ones first followed by the medium, and then shortest ones. The last angled shingle on a course should end up being about 4 in. tall or so, which will maintain the proper built-up thickness and help prevent any depressions or humps in the wall (see photo above).

## Finishing Up

I need to take a few measurements when I get close to installing the last couple of courses. Sometimes I find that the eave or soffit is out of level, so I need to adjust slightly the last few courses to look parallel and level. For example, a soffit that is out of level 1 in. over 20 ft. will be noticeable if the shingle courses remain level. By dropping the low side ⅛ in. over the final eight courses, that discrepancy can be made to disappear. Of course, when I start shingling the section on the other side of that corner, I need to make a corresponding adjustment to the courses on that side as well.

Provided that the eave or soffit is level, I can save time and create smoothly cut shingles by cutting the last two courses on a table saw. In some cases, the trim will be packed out so that the cut ends of the shingles tuck underneath. I use stainless-steel or brass nails for the exposed nailing on the last courses, and I locate the nails consistently because they will be visible.

*Oliver Govers is a roofing and siding contractor in Oakland, CA.*

# Restoring Wood with Epoxy

■ BY TOM O'BRIEN

"Give me liberty or give me death!"

**O**n March 20, 1775, Patrick Henry spoke those immortal words in a small church here in Richmond, Virginia, galvanizing the second revolutionary convention to fight for independence from England. As a carpenter living and working in Richmond's historic Church Hill district, named for the church where Patrick Henry delivered those famous words, I am constantly bombarded by history.

As a lover of history, I am both fascinated and humbled to know that many of the structures I work on have borne witness to profound events. So when I'm asked to rip out and replace a piece of weather-damaged woodwork where Robert E. Lee may have leaned a war-weary shoulder, I try to find an alternative. High-strength epoxies offer that alternative whether the wood you're fixing is 20 years old or 200 years old. With apologies to Mr. Henry, I have to say, "Give me epoxy."

## Epoxy Isn't for All Repairs

Epoxy for wood repair has been around for decades and is routinely specified on major historical-restoration projects nationwide. One manufacturer, Abatron, has been making epoxy wood-repair systems since 1959.

In the simplest terms, an epoxy repair goes something like this. After loose, crumbly wood is removed and sound wood is dried, a liquid consolidant is applied to impregnate and strengthen rot-damaged wood. Then a puttylike compound fills the gaps, replacing the missing wood.

The most common complaint I hear from people who've attempted an epoxy repair is that "the damn stuff all fell out." But after using epoxy wood-repair systems for many years, I've found that if procedures are followed carefully, epoxy is almost as wonderful as the advertisements claim. But when instructions aren't carried out carefully and dutifully, indeed, the damn stuff may "all fall out."

Epoxy does have its drawbacks. First, it's expensive, at least 50¢ per fluid ounce, and that's for large quantities. Also, in many cases the repair procedure is more complex

**Problems and Solutions.**
The photos at left show a few of the many ways that epoxy can be used to repair damaged wood. Although severely rotted wood must be removed, other damage can be strengthened and stabilized with a liquid consolidant in preparation for epoxy application.

**Fascia**

**Windowsill**

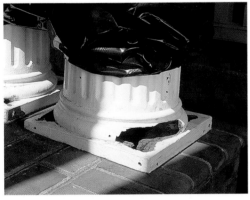

**Column base**

and time-consuming than a simple cut-and-replace job. In my opinion, some rotten-wood situations, such as framing lumber, square porch posts, and stock railing parts, are usually easier to replace than to repair with epoxy, unless removal of the damaged wood is difficult, the damage is minimal, or George Washington's initials are carved into the damaged wood.

Epoxies can also be used for major structural repairs such as rebuilding the tenons on large beams in historic timber frames. In these repairs a special epoxy that cures harder than conventional epoxy is combined with gravel, fiberglass-reinforcement rods, or metal plates to effect the repair. Structural repairs with epoxy require special skills and should be done by epoxy experts with the approval of a structural engineer.

On the other hand, nonstructural repairs such as window parts, turnings, door-jamb bottoms, and decorative cornice work as well as minor structural repairs, such as column bases, are almost always worth restoring with epoxy and require only basic skills and common tools. I try to approach each new situation with an open mind, considering both replacement and epoxy as repair options. I don't make a decision until I have seriously evaluated the damage and considered the work involved.

## Address the Cause of the Problem First

A while back, I repaired a fascia board with a small rotten area that was caused by a leaky box gutter. Despite my pleas, the homeowner never repaired the leak, and as you can imagine, the damn stuff all fell out. Regardless of whether you decide to replace or repair

rotten wood, you must eliminate the reason that the wood rotted in the first place.

Like any well-executed project, a successful epoxy repair begins with careful prep work. I start by digging out the loose, crumbling, spongy decay until I reach sound wood, which is wood that is still firmly attached to the original stock (see photo below). The sound wood may be wet, brown, and ugly, but as long as the wood doesn't crumble or break off, I can dry it out, consolidate it, and make it part of my repair.

My favorite tools for rot removal are a painter's five-in-one tool and a wire brush. A screwdriver, pocketknife, or scratch awl is also handy for digging into deep, hard-to-get-at areas. After I've gotten rid of the nasty stuff, I remove any paint around the edges of the repair site with a scraper and a heat gun. Then I blow out all the loose debris and dust with compressed air. A shop vacuum and dust brush also work well. A clean surface is one of the keys to a permanent repair.

**Getting down to sound wood.** The first step is removing loose, crumbly wood. A five-in-one painters' tool and an awl are best for digging out any wood that is not still firmly attached.

# Drying out the Damage

Excess moisture in wood keeps it from absorbing liquid consolidant, and unconsolidated wood may continue to rot over time. So the next step after removing rot is letting the remaining wood dry sufficiently.

For minor repairs such as a small rotten area in a fascia board, a day or two in dry air may be enough once the moisture source has been eliminated. Drying out larger repairs, however, such as a severely damaged column base where a great deal of moisture is trapped in the sound wood, may take more-intensive measures. If the wood is really saturated, I drill a number of ¼-in. weep holes up from the bottom and the sides to let water run out. Then I make a tent out of 6-mil black plastic and duct tape to keep bad weather away from the repair while allowing moisture to escape. I use black plastic instead of clear because it heats up more and speeds the drying process. With the plastic tent in place, all but the wettest areas should dry within a week.

If it's convenient, you can raise the plastic tent on days when no rain is predicted to allow more air circulation. If dampness remains after a week or if I have to start the repair sooner, I accelerate the drying by fanning the wood with a heat gun set on low.

When is the wood dry enough? Most manufacturers specify that the moisture content should be less than 20%. A moisture meter is the best way to measure wetness accurately, especially when it may be trapped beneath the surface; however, this device can cost a couple hundred dollars. Most of the manufacturer representatives that I talked to told me that a moisture meter is not absolutely necessary. Basically, they said to follow the Tom O'Brien moisture test: If the wood feels wet, then it's too wet.

# Consolidation Makes Rotten Wood as Hard as Stone

Simple repairs can be consolidated by merely brushing the liquid-epoxy consolidant on the surface. If the rot is extensive or cannot be easily reached with a brush or if I suspect that rot may be hidden beneath what appears to be sound wood, I drill a series of ³⁄₁₆-in. or ¼-in. holes in a random pattern about 1 in. apart, taking care to stop the drill bit just short of the bottom of the wood. I use these holes to inject consolidant deep into the damaged area so that the consolidant can soak into the soft but sound wood.

After a final cleanup to remove the sawdust from the drilling of the injection holes, it's almost time to mix the consolidant. However, before mixing any epoxy, I always sit down with a big cup of coffee and a couple of chocolate doughnuts and thoroughly reread the manufacturer's instructions. I have a three-ring binder in which I keep the manufacturer's instructions to all the epoxy products I use, along with my own detailed notes and any other literature that I've found informative. Making certain that I haven't overlooked anything in the preparation and reacquainting myself with the whole process is well worth any razzing I may receive about having to read the directions.

Epoxies have a limited working time, or pot life, so before I begin mixing, I make sure that I have all necessary tools on hand. The list includes duct tape and 6-mil plastic to protect surrounding areas from drips and runoff, 8-oz. hair-dye bottles (the kind with the nozzle top and graduations on the side) for mixing and injecting consolidant, tongue depressors or cedar shims for stirring, tin cans or plastic margarine tubs for brushing reservoirs, a variety of disposable

**TIP**

*Before mixing and applying epoxy, slip on three or four pairs of latex gloves. When one pair gets messy, simply peel it off and you'll have a fresh pair to work with.*

**Consolidant is injected into the repair.** Liquid epoxy, or consolidant, is squeezed into holes drilled around the repair. The consolidant saturates the soft wood, making it rock hard.

**Brushed-on epoxy consolidates exposed wood.** In addition to being injected, consolidant is brushed onto exposed wood until it no longer soaks into soft wood.

paintbrushes and acid brushes for applying the consolidant, and, most important, a lot of disposable latex gloves to protect my hands. For the filling stage I'll also need some small (1 sq. ft. to 2 sq. ft.) plywood scraps that I mix the filler paste on and bunches of plastic putty knives of various widths (1 in., 1½ in., 3 in., and 6 in.) for mixing and applying the epoxy filler.

After masking off my work area, I slip on three or four pairs of latex gloves. (When one pair gets messy, I simply peel it off and have

a fresh pair to work with.) I mix the consolidant by pouring the specified amounts of resin and hardener into a hair-dye bottle. Then I stir the mixture thoroughly for a full 5 minutes.

When I'm finished stirring, I pour about half of the liquid into an empty container for brushing and screw the nozzle top on my mixing bottle. Using the spout of the hair-dye bottle, I squeeze consolidant into the holes I drilled until the consolidant oozes out of the holes (see top photo). I also brush the liquid over the exposed wood (see bottom photo on facing page). Dry, damaged wood soaks up the liquid consolidant. So I continue to inject the liquid until the holes stay full, and I brush it on until the surface of the damaged wood is saturated and glossy.

## Filling the Gaps

The consolidant should have a few hours to dry, but it does not have to be fully cured before filling can begin. In fact, the best bond between the consolidated wood and the filler is achieved when the consolidant is still a little tacky. If the consolidant has fully hardened before I can fill the damaged area, I lightly sand the area or brush it with a light coat of consolidant to help the filler bond to the wood.

The mixing procedure for epoxy filler depends on the brand used. Some epoxies, such as ConServ® and West System, start as liquids. With this type of epoxy, the resin and the hardener are combined in a small paint bucket or similar container. Once the liquid has been thoroughly stirred, powdered thickening agents are added to create the filler paste.

Other types of epoxy, such as Abatron, start out as two-part pastes, which are ready to mix with no other additives (see top photo). For this type of epoxy, I combine the two parts on a mixing board and knead the mixture thoroughly for at least five minutes with a 1½-in. plastic putty knife, scraping the surface of the board often to ensure a homogeneous mixture.

After the paste is mixed, I spread the filler out on a board in a thin layer (see bottom photo). Epoxy creates heat as it cures, and heat hastens the curing process. I've been told that a large mass of curing epoxy can actually generate enough heat to catch fire. Keeping the mixture in a thin layer minimizes heat buildup and lengthens the pot life of the mixture.

**Two-part fillers are mixed on a board.** After measuring the right amount of resin and hardener paste, the two are kneaded together on a scrap of plywood.

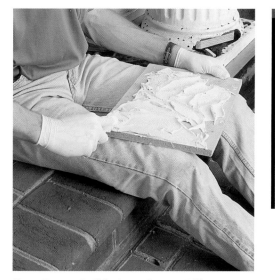

**Keep the paste thin to prolong pot life.** Epoxy cures more quickly if left in a large mass. To lengthen working time, the mixed epoxy is spread out in a thin layer before the filling begins.

*The best bond between the consolidated wood and the filler is achieved when the consolidant is still a little tacky.*

**Wood blocks save on epoxy.** To keep from using large quantities of expensive epoxy, blocks of wood are encapsulated in the epoxy to act as aggregate.

**The rest of the hole is filled by hand.** Epoxy paste is packed into the repair by hand until the entire gap is filled.

For a small repair, I usually fill the entire hole in one shot. However, large repairs have to be treated a little differently. Because of the heat-buildup problem with a large mass of curing epoxy, manufacturers recommend applying the epoxy in layers no more than 1 in. thick at any one time to avoid the risk of fire.

Large quantities of epoxy can also get pretty expensive. So to keep down the cost and to do most of the filling without risking fire or compromising structural integrity, I use small blocks of clean, dry wood as aggregate in the epoxy (see top left photo). First I cut the blocks roughly to size and dry-fit them, making sure there is at least ¼ in. of space on all sides of the block for the epoxy. Next I coat the insides of the holes with plenty of epoxy paste. After buttering the blocks on all sides with the paste, I press them into the holes. The epoxy should ooze out on all sides of the blocks to ensure that no air pockets are left behind the blocks.

**All dressed up.** The same poly tent that is used to keep moisture away from the repair while it dries also protects the epoxy from weather while it cures.

**A homemade tool for re-creating odd shapes.** Irregular profiles such as the fluted bottoms on these columns can be molded using a plastic putty knife with the shape cut into the blade.

I fill any remaining spaces with epoxy (see bottom left photo on facing page) and let the repair cure overnight under the poly tent (see bottom right photo on facing page).

## Cured Epoxy Can Be Worked like Wood

If the area I'm repairing has an irregular shape, such as the fluted base of a round column, I tool the final layers of paste with a plastic putty knife cut to fit the contour of the irregular shape (see photo above). Once the epoxy has cured, it is delightfully easy to work (see photo at right). I use a Stanley Surform® pocket plane and a wide, sharp wood chisel to do the rough-shaping. Then I sand the repair smooth with 100-grit paper, using a random-orbit sander wherever possible.

**Woodworking tools also handle epoxy.** Cured epoxy can be worked like wood. A chisel and a Surform plane rough out the shape, and sandpaper does the smoothing.

With most repairs I usually have to touch up a few spots with a skim coat of epoxy (see photo at left). To achieve a smooth finish—and to reduce the amount of sanding—I mix a slightly more fluid batch of epoxy. Manufacturers have recommendations for adjusting the consistency in their instructions. Depending on the product, this is done by reducing the amount of thickeners, by adding a thinner or consolidant, or by adding a specially designed liquid epoxy. When I'm in a rush, I fan the skim-coated areas with a heat gun set at low temperature, taking care not to cause the skim coat to blister. A heat gun can make the epoxy skim coat cure in 10 minutes to 15 minutes, and a quick pass with the sander blends it in with

**A skim coat touches up small voids.** Epoxy filler adheres well to itself, so any small areas left from the initial fill can be skim-coated and blended into the repair.

**Primer protects the epoxy.** As a final step before painting, a coat of primer is applied to seal the repair against moisture as well as against UV radiation, which can damage some epoxies.

the rest of the repair. As a final step before the painter arrives, I give the repaired area a coat of a high-quality exterior primer (see bottom photo on facing page). Priming and painting not only protect the repair from the weather but also fend off the ultraviolet radiation that some epoxies are vulnerable to.

# Keep an Eye on the Repaired Areas

Although I consider a careful epoxy repair to be permanent, that doesn't mean it can be forgotten. The fact that the wood failed in the first place means it may be susceptible to damage in the future. For this reason I always urge homeowners to monitor the repair as well as neighboring areas as part of their regular maintenance schedule, along with keeping their rain gutters cleaned out, painting, and doing other chores that keep moisture from deteriorating wood.

Maintenance should be twofold. First, make sure that whatever water leakage caused the wood to rot in the first place does not recur. Second, check the repair once or twice a year for soft areas that may develop where the consolidant may not have reached or where rot may have gone undetected. If discovered early, repairs can be made easily.

---

*Tom O'Brien* is a associate editor of Fine Homebuilding *magazine.*

## Sources

There are too many manufacturers to include in this limited space, but here is a list of the epoxy makers I've had experience with.

**WOOD-RESTORATION EPOXIES**

**Abatron, Inc.**
(800) 445-1754
www.abatron.com

Abatron is the oldest and largest maker of epoxies for wood repair. Available only by mail order, their products are easy to mix and tool, and they come in convenient, easy-to-use packaging. Starter kits are available in a variety of quantities.

**Advanced Repair Technology**
(607) 264-9040
www.advancedrepair.com

Instead of using consolidation, Advanced Repair's system requires that all of the damaged wood be removed before rebuilding. A unique mixing system that uses twin (caulking-style) tubes takes some of the guesswork out of epoxy preparation.

**ConServ Epoxies**
(973) 579-1112
www.conservepoxy.com

ConServ is a much smaller operation than Abatron, but their products come with good instructions (a 65-page application manual is available for $10). Technical support is good, and the company's owner, Bruce Whipple, will talk you through the procedure. ConServ takes longer to cure but after curing stays more flexible than most epoxies.

ConServ is available only by mail order but comes in less convenient packaging. The mixing procedures are also more involved.

**MARINE EPOXIES**

**West System Epoxy**
(989) 684-7286
www.westsystem.com

West System is intended mainly for boat repair, but it's also marketed for repairing wood rot. It's available at marine-supply centers nationwide and comes with an excellent technical manual. I'm told that Gougeon Brothers are renowned for their terrific technical support.

West System also uses no consolidant, and the 5:1 mixing ratio can be complicated.

**Smith and Co.**
(800) 234-0330

Smith is another marine epoxy. I bought some but haven't had a chance to try it yet. It's also marketed for wood repair.

**OTHER PRODUCTS**

**Stanley Tool Works**
(800) 648-7654
*Surform pocket plane*

**Thompson-Minwax® Co., Inc.**
(800) 523-9299
www.minwax.com

The Minwax product is actually a polyester rather than an epoxy. It looks and acts much like auto-body Bondo® and is available in hardware stores and home centers nationwide. However, their instructions are skimpy, and they offer no technical support.

**John Leeke**
(207) 773-2306
johnleeke@
historichomeworks.com

Leeke is a preservation consultant. He publishes a series of practical restoration reports, including one called "Epoxy Repairs for Exterior Wood" (10 pp., $9.95), which is filled with many helpful tips.

# Removing Exterior Paint

■ BY JON TOBEY

When I was a kid, I never had to look for work. From the time I was 12 until I graduated from college, my summer job was scraping and painting my parents' sprawling, 175-year-old New Hampshire farmhouse. Every few years, I'd complete a lap, only to have to start over again. After I became a professional painter, the value of doing a job once and doing it right sank in. Over the past 15 years, I've refined the methods and the materials that I use to ensure a long-lasting paint surface. These days, I can confidently say that if it's done correctly, a good paint job should last 20 years.

## Sometimes You Have to Take It All Off

A long-lasting paint job depends on high–quality paint (and plenty of it), but prep work is even more important. Before repainting, most conscientious painters give a house a thorough hand-scraping and pressure washing. That's fine as long as all the existing layers of paint are securely bonded to each other and to the substrate. On my parents' house, like many of the houses I've worked on, that was not the case.

I'm not trying to say that all houses must be stripped bare before repainting. In my experience, unless they suffer from specific moisture problems, newer houses (up to 35 years old) are less troublesome than older houses. If a house was primed and painted correctly when new, especially if it was painted with a high-quality, flexible latex, the paint finish should last a long time and should be easy to recoat when it wears out.

Repainting is similar to reroofing, in that you can pile only so many layers on top of one another before it's time for a tearoff. At the time I started my summer job, my parents' house already had six or seven distinct layers of paint on it. Oil paint is much less flexible than latex, and its ability to expand and contract along with the wood diminishes with each new layer. On older houses, unless the finish is thin or in good shape and well protected from the sun (expansion and contraction are most extreme on the southern exposure), I generally advise homeowners to bite the bullet and allow me to remove all the old oil-based paint before I begin repainting with latex.

When you have to take it all off, power tools and chemical strippers make quick work of a nasty job.

**Chemicals are the last resort.** After stripping most of the clapboard surfaces with the Paint Shaver, the author uses a nontoxic chemical paint remover to finish up in the corners. (Note: Photo taken before the author discovered the Metabo Paint Remover.)

# Mechanical Scrapers Get the Job Done Quickly

It's easy to make the argument for removing failed paint, unless you have to come up with the cash to pay for it or do the work yourself. Over the years, I've tried all the traditional methods for removing paint and found serious drawbacks with each:

- Hand scraping is slow and grueling.
- Power sanding is slow and messy.
- Heat is slow and dangerous (even low-temperature heat guns can start a fire).
- Chemical paint removers are slow and expensive (most are also toxic).

It's no wonder that most painters would rather convince themselves (and their customers) that a quick scrape and wash is all the prep work a house really needs.

Until a few years ago, I would turn down a job if I determined that the paint had to come off. Then I discovered the latest generation of mechanical paint removers. These power tools don't make the work any more pleasant, but they get the misery over and done with quickly. No power tool is capable of stripping all the paint from all the nooks and crannies, so my strategy is to remove the paint mechanically wherever I can, then use a chemical paint stripper on the rest.

The workhorse in my arsenal is a tool called the Paint Shaver® from American International Tool. A modified mini-grinder, the Paint Shaver makes quick work of wide, flat painted surfaces such as clapboards and shakes. The only drawback to the Paint Shaver is that its cutterhead (see left photo on facing page) isn't much good for corners and tight spaces. Until recently, whatever paint I couldn't reach with the Paint Shaver I'd have to attack with chemicals.

I recently tried out some other paint-removal tools. One tool I really like is the Metabo® Lf714S Paint Remover from Metabo Corp. The Metabo is nowhere near

# Mechanical Scrapers Tackle a Tough Job

Specially designed power strippers such as American International Tool's Paint Shaver and Metabo's Paint Remover quickly remove multiple layers of old paint and collect the debris for disposal.

Triangular carbide blades cut on two surfaces at a time to remove paint from edges and faces of clapboards.

Screws control depth of cut on edges.

Cutterhead

Blades can be rotated to use all six cutting edges.

Steel guard controls the depth of cut on clapboard face.

Dust-collection shroud contains debris.

**Paint Shaver, American International Tool**

The powerful Paint Shaver makes quick work of wide, flat surfaces.

**Metabo Lf714S Paint Remover**

The Metabo Paint Remover, thanks to its smaller footprint and tighter cutting radius, cleans up in the corners.

Lead-based paint was outlawed in 1978, so any house built before that time could contain lead-based paint. The best way to address the problem is to have the house professionally tested to determine if, and where, lead-based paint is present. (Contact your local public-health organization for information on lead-inspection services in your area.) You also can pick up a lead-paint test kit at a hardware or paint store. These kits let you know if lead is present in paint; however, unless you test every side of the house, it is possible to get a false-negative result.

If I'm working on a house that's more than 30 years old and I don't know for a fact that it's lead-free, my rule of thumb is to assume that the paint is lead-based. Even though my dust-collection equipment captures 95% of paint debris, on these jobs, I always wear a full-face respirator and a Tyvek suit; I also tape the openings at the neck, wrists and ankles. (Note: The section of the house shown in these photos did not contain lead paint.) When I'm done for the day, I undress outside my house (behind a bush); then my clothes immediately go into the wash, and I hit the shower.

as fast as the Paint Shaver, but its small footprint makes it the perfect follow-up tool for getting into corners and cleaning out the small stuff.

## Protection Measures Come First

State-of-the-art power tools may make paint removal faster, but no less exhausting, especially if you're working from an extension ladder the whole time. When I'm doing a large job (unlike the small one I did for the photos), I prefer to put up staging—either pump jacks or pipe scaffolding—and wrap the entire house. Doing this job on staging is much easier on the body because you can work directly in front of your torso and apply more power and finesse with less effort. The rental cost of the staging is more than offset by the increase in productivity and morale.

Once the work site is set up, it's time to run out the tarps. Even though I use vacuum attachments that suck up 95% of the debris, I always cover the ground 6 ft. to 8 ft. out from the house with heavy-duty painters' drop cloths.

If I'm using chemicals or if I suspect I might be removing lead-based paint, I use 3-mil painters' plastic instead of drop cloths. At the end of each day, I can carefully roll up the plastic to dispose of debris. In some locales, this method is sufficient to let the debris be put out for regular trash pickup, but check local disposal regulations.

## Scraper Removes Paint from Two Surfaces at Once

Before I start grinding paint, I need to make sure that nail heads are set below the surface of the wood; otherwise, they'll wreak havoc on the scraper blades. At $49 a set, replacement blades aren't cheap; but if I'm careful,

# Why Good Paints Go Bad

Chalking just requires a good cleaning.

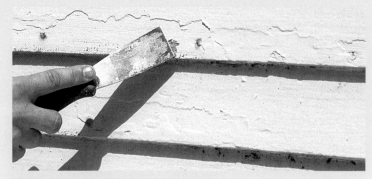

Sharp edges need scraping and sanding.

Alligatoring means it all has to come off.

Peeling in sheets points to hidden moisture problems.

Paint failure is a broad topic worth an article to itself. As a painting contractor, I examine failed paint to determine what prep work I have to do to make sure my paint stays put.

Chalking is a sign that good paint has reached the end of its useful life. Even if the underlying paint surface is sound, the "chalk" must be washed away, or the new paint won't stick.

Sharp edges where an older layer of paint slices through a newer one point to insufficient hand scraping. Usually this problem can be fixed by scrapping the loose paint and feathering the edges with a palm sander.

Other paint failures are more serious. Cross-grain cracking, or alligatoring, arises from too many coats of paint. Alligatoring is a sign that paint needs to come off.

Paint peeling away in sheets is often a sign of underlying moisture problems that must be corrected before repainting can begin.

A high-quality latex over oil can be a problem. The latex bonds with the oil underneath, and when the latex expands and contracts due to temperature differentials, it pulls the more brittle oil paint right off the surface. For this reason, it's often best to remove all the oil paint from southern exposures on older houses before repainting with latex.

a full set may last an entire painting season. If the job is completely staged, I run around and hit all the nails at one time; otherwise, I carry a hammer and set any protruding nails as I go.

When I'm stripping paint, I don't want debris to rain down the wall and cling to lower surfaces, so I always start at the bottom and work my way up. That way, I'm always working into a clean area. What makes the Paint Shaver such an effective workhorse is that its multisided carbide blades are mounted in the cutterhead so that they remove paint simultaneously from the face of one clapboard and the bottom edge of the one above. A U-shaped steel guard that surrounds the blades controls the depth of cut. To reduce the likelihood of damaging fragile clapboards, I adjust the tool for a shallow cut and make several passes over the wall surface.

Even with blades set for a shallow cut, if you've never used one of these tools before, you can chew up a lot of good wood before you get the hang of things. I always make sure new employees get plenty of practice on an inconspicuous spot before I turn them loose on the home's front entrance.

When I'm operating the Paint Shaver, I make sure that the steel guard (see left photo on p. 145) that surrounds the cutterhead remains flat on the painted surface while I move the tool from right to left in a series of narrow (shoulder-width) passes. To ensure stability, I keep one hand on the auxiliary handle and guide the body of the scraper with my other hand (see top photo in inset box on p. 145).

## Debris Is Collected at the Source

If you've never used the Paint Shaver before, you should practice with the dust-collector removed (and wear a full-face respirator) because it's much easier to see what you're doing. Once you've gotten used to working partly by feel rather than entirely by sight, the value of the dust collector becomes apparent. My nonscientific opinion is that it picks up more than 95% of the debris, which makes it useful for any job but essential wherever lead paint might be involved. Even the best dust collector is useless without a heavy-duty shop vacuum and a hose that's as long as the extension cord. I use twist ties to keep the extension cord and the vacuum hose together (see photo on p. 143).

The Paint Shaver isn't much good in corners, so now I just give them a light kiss and come back later with the Metabo. Although its manufacturer markets the Metabo Lf714S as a full-scale paint remover, it doesn't have nearly the power of the Paint Shaver, so it would take much longer to do a whole house. What it does have is finesse. The Metabo has a long, flat sole plate that makes it as stable in operation as a palm sander. Its depth adjustment is even finer than that of the Paint Shaver, and what's more, the cutting heads can be exposed on both sides as well as on the front (see bottom left photo in inset box on p. 145). This setup makes it easy to reach deep into corners (see bottom right photo in inset box on p. 145).

## Chemical Stripper Cleans out the Nooks and Crannies

After I've hit each corner with the Metabo paint remover, there's usually a silver-dollar-size wedge of paint that needs attention. To get the last bit of paint out of the corners, and (more important) to remove paint from trim details, I use chemicals. About a year ago, I discovered a new product called Back to Nature Multi Strip™ that consistently removes multiple layers of paint (latex as well as oil) with one application. It's also relatively safe. Whereas most paint strippers are extremely caustic, Multi Strip is waterbased and requires no special neutralizing agents.

Despite this product's benign nature, I still put on rubber gloves, long sleeves, and long pants before I begin the application process. Using a disposable 4-in. brush, I apply the stripper, a semipaste, in a thick, even coat, then wait for it to take effect. Air temperature, paint thickness, and types of paint all influence the effectiveness of the chemical stripper. All I can do at this point is watch the area and look for signs of bubbling and discoloration, occasionally testing a small section until I'm sure that the stripper has done all the work that it's going to. In my experience, this process can take anywhere from 4 hours to 24 hours. Then I scrape the surface and apply a second coat, if necessary.

## Trim Surfaces Need the Least Attention

Once all the siding has been stripped, I often stop there. Trim surfaces rarely undergo expansion and contraction to the same degree that wall surfaces do, so even if it does not look better, the trim paint is usually in better shape. Most paint failure on trim is localized (rather than systemic) and does not require removal to bare wood. In most cases, I simply hand-scrape the loose paint, then sand the edges. For hand-scraping trim elements, especially complex profiles, I use ProPrep® scrapers. These well-balanced tools come with a wide variety of interchangeable blade profiles.

**To remove dry or chemically softened paint from the nooks and crannies of old trim, the author swaps between interchangeable blade profiles in Spectrum Tools' ProPrep scraper.**

**Getting rid of the fuzz.**
After the paint is gone, a palm sander loaded with 40-grit sandpaper prepares an old softwood surface for primer.

### Sources

**American International Tool**
(800) 932-5872
www.aittool.com

**Back to Nature**
(800) 211-5175
www.ibacktonature.com
*Multi strip*

**Metabo Corp.**
(800) 638-2264
www.metabousa.com

**Spectrum Tools**
(800) 229-2233
www.razortools.com

*ProPrep Scrapers*

When I encounter massive paint failure on trim surfaces, I'm more dependent on chemical paint removers than when I'm working on wall surfaces. Fortunately, the combination of a sharp blade, the right profile, and some effective chemicals makes this job faster and easier than it was in the past.

Mechanical paint removers, by design, are rough-cutting tools. To prepare a surface properly for paint, most tool manufacturers recommend power-sanding the surface after the paint has been removed. Unless the homeowner insists on perfectly smooth surfaces, I prefer to minimize this step. In fact, I've found that the "tooth" left behind by the paint remover aids significantly in helping the new paint bond to the house, so I take pains not to oversand.

However, after the paint is removed from old softwoods, such as white pine, the surface is left with a furry nap that must be taken down (see photo above). In these cases, I quickly go over the entire surface with a palm sander.

Before priming, I lightly pressure-wash the bare wood to remove dust and chemicals. Whenever lead paint is involved, I make sure to capture the dust from the house and not let it wash into the soil. A light wash can be contained using 2x4s and plastic drop cloths. After the water has evaporated, the plastic is carefully bundled up for disposal.

*Jon Tobey is a painting contractor and frequent contributor to* Fine Homebuilding *magazine. He lives in Monroe, WA.*

# CREDITS

p. iii: Photo by Andrew Wormer, courtesy of *Fine Homebuilding,* © The Taunton Press, Inc.

Table of contents: Photos on p. iv (left) by Roe A. Osborn, courtesy of *Fine Homebuilding,* © The Taunton Press, Inc.; p. iv (right) © Steven Snyder; p. v (left) by Roe A. Osborn, courtesy of *Fine Homebuilding,* © The Taunton Press, Inc.; p. v (center) by Charles Miller, courtesy of *Fine Homebuilding,* © The Taunton Press, Inc.; p. v (right) by Andrew Wormer, courtesy of *Fine Homebuilding,* © The Taunton Press, Inc.

p. 5: Running Exterior Trim by Robert Weatherall, issue 136. Photos by Roe A. Osborn, courtesy of *Fine Homebuilding,* © The Taunton Press, Inc. Illustrations by Rick Daskam, © The Taunton Press, Inc.

p. 16: Building an Ornate Cornice by Terry Murphy, issue 86. Photos on pp. 17 and 23 (bottom left) by Charles Miller, courtesy of *Fine Homebuilding,* © The Taunton Press, Inc.; pp. 18–21, 23 (top and bottom right) © Terry Murphy. Illustrations © Bob La Pointe.

p. 24: Exterior-Trim Details That Last by John Michael Davis, issue 141. Photos by Tom O'Brien, courtesy of *Fine Homebuilding,* © The Taunton Press, Inc. Illustrations by Christopher Clapp, © The Taunton Press, Inc.

p. 34: Synthetic Stucco by Steve Culpepper, issue 105. Photos on pp. 34, 36–40 by Steve Culpepper, courtesy of *Fine Homebuilding,* © The Taunton Press, Inc.; pp. 35, 43 by Scott Phillips, courtesy of *Fine Homebuilding,* © The Taunton Press, Inc. Illustrations by Heather Lambert, © The Taunton Press, Inc.

p. 45: Flashing Brick-Veneer Walls by Douglas R. Stieve, issue 142. Photos by Roe A. Osborn, courtesy of *Fine Homebuilding,* © The Taunton Press, Inc. Illustrations by Don Mannes, © The Taunton Press, Inc.

P. 53: Installing Manufactured-Stone Veneer by Vladimir Popovac, issue 130. Photos by Roe A. Osborn, courtesy of *Fine Homebuilding,* © The Taunton Press, Inc., except p. 59 photos courtesy of Cultured Stone.

p. 62: Laying up Stone Veneer by Steven Snyder, issue 91. Photos © Steven Snyder. Illustration by Dan Thornton, © The Taunton Press, Inc.

p. 72: Vinyl Siding Done Right by Mike Guertin, issue 149. Photos by Tom O'Brien and David Ericson, courtesy of *Fine Homebuilding,* © The Taunton Press, Inc.

p. 81: Installing Wood Clapboards by Rick Arnold and Mike Guertin, issue 112. Photos by Andy Engel, courtesy of *Fine Homebuilding,* © The Taunton Press, Inc. Illustrations by Rick Daskum, © The Taunton Press, Inc.

p. 92: Working with Fiber-Cement Siding by John La Torre, Jr., issue 140. Photos by Roe A. Osborn, courtesy of *Fine Homebuilding,* © The Taunton Press, Inc., except p. 101 photo © John La Torre, Jr.; p. 94 siding samples courtesy of Cemplank.

p. 102: Installing Horizontal Wood Siding by Felix Marti, issue 96. Photos by Charles Miller, courtesy of *Fine Homebuilding,* © The Taunton Press, Inc. Illustrations by Dan Thornton, © The Taunton Press, Inc.

p. 114: Rain-Screen Walls: A Better Way to Install Siding by Mark Averill Snyder, issue 137. Photos © Mark Averill Snyder. Illustrations by Rick Daskam, © The Taunton Press, Inc.

p. 122: Siding with Cedar Shingles by Oliver Govers, issue 102. Photos by Andrew Wormer, courtesy of *Fine Homebuilding,* © The Taunton Press, Inc.

p. 132: Restoring Wood with Epoxy by Tom O'Brien, issue 107. Photos by Roe A. Osborn, courtesy of *Fine Homebuilding,* © The Taunton Press, Inc.

p. 142: Removing Exterior Paint by Jon Tobey, issue 148. Photos by Tom O'Brien, courtesy of *Fine Homebuilding,* © The Taunton Press, Inc.

# INDEX

Index note: page references in italics indicate a photograph; references in bold indicate a drawing.

## R

R. J. Kenney Associates, 41, 44
Rain-screen walls:
    airspace and capillary break in, 115–16, *116*, **117**, 118, *118*
    back priming of, 119–20
    clapboard siding application to, **117**, 120–21
    flashing applications in, 116, *116*, **117**, 118, *118*, 119
    membranes in, 116, *116*, **117**
Rake trim:
    finishing gables with, 14, **15**
    vinyl siding and, 74, *74*
Red cedar, exterior trim using, 4
Restoring wood with epoxy:
    appropriate jobs for epoxy, 132–33, *133*, 134
    of column bases, *133*
    drying out the damage before, 135
    filling the gaps in, 137, *137*, 138, *138*, 139, *139*
    identifying causes of problems before, 134
    monitoring repaired areas with, 141
    removing rot and crumbly wood in, 134, *134*
    using liquid-epoxy consolidant in, 135–36, *136*, 137
    working cured epoxy like wood, 139, *139*, 140, *140*, 141
Rolf, "Sunny" Jim, 17, *18*, 19, 22
Rot acceleration weather, 25

## S

Siding:
    back-priming of, 115
    synthetic stucco compared with, 40, *40*
    *See also* Cedar shingles
    *See also* Fiber-cement siding
    *See also* Horizontal wood siding
    *See also* Rain-screen walls
    *See also* Vinyl siding
    *See also* Wood clapboards
Soffits:
    exterior trim eave details for, 6, **6**, *7*, 8, *8*, *9*, *10*
    flashing of, 11, *12*
    preparation for vinyl siding and, 74, *74*
    soffit-to-fascia joints with, 8
    working with water table, *10*, 11, *12*
Staggered butt joints, soffit-to-fascia joints with, 8
Stone veneer:
    date stones with, 70–71, *71*
    disguising corner returns of, 67–68, **68**
    foundations for, 64, *65*
    laying up, 62–71
    mortar for, 62, 68
    point stone in, 70–71, *71*
    reference lines for, 64–65, *65*
    setting windows and wall ties with, 67, *67*, 70, *70*
    using salvaged stone, 62, *63*, 64, *64*
    ventilation in walls of, 69, *69*
    *See also* Manufactured stone veneer

Story poles:
    history of synthetic stucco with, 35
    in horizontal wood siding applications, *103*, 108, *108*
    in wood clapboard layout, **84,** 85
Stucco:
    with building papers, 121
    rain-wall screens and, 121
Synthetic stucco:
    architectural elements with, *37*
    around the windows, 41, **41**, 42, **42**
    compared with ordinary siding, 40, *40*
    expert research of, 40, *40*, 41
    exterior insulation and finish systems of (EIFS), 34–44
    finish coat of, 39, *39*, 40
    history of, 35
    installation of, 35, *35*, 36, *36*, 38, *38*
    points for water intrusion in, 36–37
    removal of, *34*, 35
    water intrusion in, 35, *35*, 36, *36*, 37
    watershed designs of, 42–43, *43,* 44, **44**

## T

Tannins, from cedar siding, *114*, 115
Tar paper:
    in brick-veneer walls, 47, 49
    with cedar shingles, 126, *126*
    in rain-screen walls, 115, 118, *118*, 119, *119*
    splines with wood clapboard, 87–88, *88*, *118*
    vinyl siding with, 72, 79
    in water drainage systems, **46,** 47, 49
Trim:
    caulking of, *24*, 25, **25**, 26, *30*, 33, *33*
    in exterior trims, 4–15
    historic and archaeological research of, 16–17, *17*, *18*, *18*
    for ornate cornices, 18, **19**
    poor-man's crown in, **6,** 8
    in rake trims for gables, 14, **15**
    removing paint from, 149–50, *150*
    ventilation in, **28,** 29, *31*
    vinyl trim for windows, *78*
    wood for exterior trim, 4

## V

Ventilation:
    in exterior trim, **28,** 29, *31*
    integrating in frieze boards, *10*, 11
    in rain-screen walls, *120*
    in traditional boxed eaves, 6, **6**
    in walls of stone veneer, 69, *69*